Curing Lives

Makoto Nishi

Curing Lives

Surviving the HIV Epidemic in Ethiopia

Makoto Nishi
School of Integrated Arts and Sciences
Hiroshima University
Hiroshima, Japan

ISBN 978-981-99-1830-0   ISBN 978-981-99-1831-7   (eBook)
https://doi.org/10.1007/978-981-99-1831-7

Cover illustration: © Melisa Hasan

This Palgrave Macmillan imprint is published by the registered company Springer Nature Singapore Pte Ltd.
The registered company address is: 152 Beach Road, #21-01/04 Gateway East, Singapore 189721, Singapore

for Betty

Preface

This book is about life during the HIV epidemic in Ethiopia. I am writing this book to ask why and how the global effort to achieve universal HIV treatment has shifted away from its initial focus on the excessive human suffering precipitated by the epidemic. When antiretroviral drugs became available in Ethiopia, they emerged as powerful agents of change: they not only cured individuals but also helped people to overcome their fear of—and break the silence around—AIDS while healing the social ruptures caused by the epidemic. Nevertheless, as I argue in this book, the same agents seem to have silently reversed these changes over the past decade. These reversals have dissolved connections, reincurred invisible social fissures, and allowed a large majority of people to stay indifferent to the suffering of individuals whose lives remain vulnerable under the current treatment regime.

This is how the pharmaceuticalization of HIV care unfolded in Ethiopia. Pharmaceuticalization is the term used to describe what happens when pharmaceutical agents such as pills are introduced to replace or alter existing healthcare methods and practices. The obvious consequence of introducing antiretroviral therapy (ART) in Ethiopia and elsewhere in Africa is that it saved millions of people from imminent death. However, less apparent is how the pharmaceuticalization of HIV care interplayed with the existing forms of inequalities and exclusions, reinforcing the vulnerable state of lives of some people in African societies.

It requires a "biopolitical" perspective to make sense of such a process simultaneously involving the biological, epidemiological, social, and political aspects. In each society, pharmaceuticalization accompanies a particular set of entitlements to care and resources that are conceded or refused under a health governance system. Medical anthropologists and sociologists refer to such entitlements as therapeutic citizenship. Where health policies are guided by the economic value of intervention instead of the concern over the individual state of life, such entitlements are likely to be shaped and enacted in a way that differentiates human lives as worthy or unworthy of investment. This process is similar to what Michelle Murphy (2017) referred to as the "economization of life" in her book concerning population and reproductive health intervention in the twentieth century.

In Ethiopia, where ART has been offered to its citizens free of charge since 2005, those regarded as unworthy of investment are often those whose problems are too complicated to solve with antiretroviral drugs. In a provincial town where I conducted my fieldwork between 2007 and 2015, such individuals were subject to the practice of "triage," which effectively removed them from the scope of economic and healthcare support that would have helped them reclaim their lives disrupted by HIV and other illnesses, economic distress, and social exclusion. However, that is not to say that alternative forms of care and social support that give place to culturally shared values and meanings of life did not take shape in Ethiopian society. In this book, I use the term "curing life" to represent the actions and engagements that resist the dominant mode of health governance that promotes population health without regard to its consequences over the often complicated states of life.

What precisely are the conditions that have promoted or resisted the economization of life during the HIV epidemic in Ethiopia? This concern has moved me to address three interrelated questions in this book. The first question concerns the scope of HIV intervention over the past decade: What rationales have shaped and driven the "universal treatment" strategy as the primary method for the global fight against HIV? What conditions enabled the rapid rollout of free ART in Ethiopia, where health infrastructure was extremely scarce? The second question concerns this strategy's impact on the lives of Ethiopians struggling to survive the epidemic. This book focuses on one particular woman—someone who has sought to "cure" her own life and the lives of those who have been associated with her. Her concern for "curing lives" has led me to ask the third question: What, in the first place, might living a moral life mean to a woman in contemporary Ethiopia? Furthermore, what forms of moral subjectivity resist the economization of life, particularly when the latter is deeply entangled in the cultural and epidemiological realities of its own time and space?

This book consists of three parts that correspond to the questions outlined above. The first part demonstrates how the rollout of ART in Africa triggered a shift in the moral rationale for global health interventions. Using a rich trove of evidence extracted from the ART experience in Africa, epidemiologists devised a "universal treatment" strategy that utilized antiretrovirals as a powerful tool for eliminating viral transmission. In Ethiopia, the strategy was materialized as one of the most coordinated national ART owing to the solid domestic political leadership supported by relatively generous international funding.

However, as health policymakers seeking cost-effective interventions embraced this strategy, the push for universal treatment began to lose its humanitarian moorings. With the rollout of low-cost antiretrovirals in Africa, the humanitarian case for universal treatment has been appropriated by a utilitarian emphasis on achieving the health of populations at a lower cost. A resulting trend is the defunding of non-pharmaceutical interventions such as home-based care activities in Ethiopia as well as elsewhere in Africa.

The second part offers a closer look at some of the effects of this unprecedented "experiment" in ensuring a healthier life for Africans. It traces, through the eyes of an Ethiopian woman whose name is Meseret, the course of the HIV movement in Ethiopia since 1997. Meseret's recounting of her life, against a backdrop of the nationwide fight against the epidemic, points to the ambivalent landscape of therapeutic citizenship in Ethiopia. Meseret's transformation coincided with the growth of the national HIV movement, which installed solidarity among people living with HIV and gave them a voice. The story shows how an isolated village woman who lost her husband to AIDS was able to reestablish her life as a member of the emerging HIV movement in Ethiopia. With members of Fana, the association of HIV-positive people Meseret established in her home province, she attempted to create a moral economy of survival in which their stories are shared and their needs articulated. This was supposed to be the space where they started the struggle to make themselves visible to society at large.

However, in contemporary Ethiopia, HIV interventions are no longer focused on addressing the realities of living with the virus. The multiple burdens borne by individuals with HIV in Ethiopia are being rendered ever more invisible within the wider landscape of therapeutic citizenship.

The last part addresses modes of engagement that resist the dominant form of governance as a manifestation of the economization of life. It concerns Meseret, as well as Asha, a traditional birth attendant in Addis Ababa, the capital of Ethiopia. While the details and trajectories of their lives differ, both women claim to have "cured some lives" through their engagements with other people's sufferings. Questions on morality, prompted by both women's concern for "curing lives," guide the discussion in this section.

This part elucidates some aspects of their lives and engagements as embedded in but not defined by local moral perspectives. It conveys accounts of local interpretations and reinterpretations of ideas concerning care, gender roles, and sociality in the wake of the HIV epidemic. Furthermore, those forms of subjectivity that manage to resist the economization of life, particularly those deeply entangled in the cultural and epidemiological realities of their own time and space, will be discussed.

This book will invite readers to visit Ethiopia several times within similar time frames covering crucial moments of the formation and transformation of HIV care, each time with different interlocutors: first with health policymakers, second with Meseret as a figure in Ethiopia's HIV movement, and then with Meseret and Asha as women who are concerned about culturally informed values and meanings of life. Each story will inform readers how the suffering of those whose lives were disrupted by HIV was rendered invisible, and the actions to cure the lives of those sufferers were marginalized.

However, at the same time, these stories point to what Didier Fassin (2009) referred to as "another politics of life," which engages the experience of individual human beings and gives place to cultural interpretations and moral decisions. In this sphere of actions and engagements, primary attention is directed to "life which is lived through a body" and "as a society" (Fassin, 2009, p. 48), unlike in the dominant biopolitics of health that focuses on populations and prioritizes economic values. In Ethiopia, Meseret and Asha are among the women who cherish the spheres of relationships left unattended by people in power. That they were concerned about culturally bound values does not mean they were submissive to the existing cultural norms. Instead, their commitment to curing the lives of others was underpinned by the "culture of defiance" or their struggle to speak and act against existing gendered norms and patriarchal claims.

Hiroshima, Japan   Makoto Nishi

References

Fassin, D. (2007). *When bodies remember: Experiences and politics of AIDS in South Africa*. University of California Press.

Murphy, M. (2017). *The economization of life*. Duke University Press.

Acknowledgements

I am profoundly indebted to Meseret Gebre for her long-standing engagement with the research that led me to create this book. Words can't describe how grateful I am to her for being kind and thoughtful all the time. I want to express my gratitude to the staff and members of Fana for sharing their stories and ideas that truly helped me to figure out some effects of the HIV interventions in Ethiopia. Many thanks to Meseret's beloved children, Shimalis and Bethlehem Habte. Your help and hospitality comforted me during my stay in Welkite. Meseret's relatives and neighbors in her home village were also very kind to me during my stay there. I owe my work to the late Asha Bule, whose stories always fascinated and inspired me. I thank her sons, Ibrahim Aman and Husein Aman, for their unsparing friendship and helpfulness. Another son of hers, Abdo Husein, facilitated my research in Silte and Sidama Zones during the late 1990s. I would like to offer my deepest gratitude to distinguished researchers at Addis Ababa University: Gebre Yntiso and Mamo Hebo at the College of Social Sciences and Dilu Shaleka at the College of Development Studies for their support and advice on my research and Haregwein Kebbede at the Institute of Language Studies for her resourceful Amharic lectures.

My research has been greatly enriched through my conversations with medical anthropologists in Japan. I sincerely appreciate Junko Kitanaka, Mohacsi Gergely, Akinori Hamada, and Yosuke Shimazono for their insightful advice, sympathy, and encouragement. Their intellectual stimulation has been genuinely formative of my research questions and arguments in this book. My research on the HIV epidemic in Ethiopia started during my days at the Center for Southeast Asian Studies, Kyoto University. I owe a debt of gratitude to Kaoru Sugihara, Akio Tanabe, Yoko Hayami, Shuhei Kimura, and Takahiro Sato. Their advice and support were indispensable in shaping my inquiry, which turned out to encompass power, technology, and care. Taizo Wada helped me make sense of some epidemiological research methods with his knowledge of public health and experience as a physician. I am grateful to professors at the Graduate School of Asian and African Area Studies, the Center for African Area Studies, and the Graduate School of Advanced Integrated Studies in Human Survivability, Kyoto University, for supporting my research.

Anthony Simpson, who visited Hiroshima in December 2022, read my manuscript and provided thoughtful comments, which were truly helpful in consolidating and clarifying my ideas presented in this book. Our conversations in Hiroshima offered me insights into a number of issues, including how the values of manhood and womanhood interplayed with the HIV epidemic in different parts of Africa. I am grateful to my colleagues at the Anthropological Institute of Hiroshima (https://taihi.org), Hiroshima University: Hirokazu Miyazaki, Koki Seki, Itaru Nagasaka, Ken Matsushima, Moe Nakazora, Chihiro Nakayashiki, and Mariko Yoshida for their constant support and intellectual input.

Without the support, patience, and understanding of my family, I have not been able to continue my research that involves regular trips to Ethiopia. I am particularly indebted to my mother, Kayoko, and my parents-in-law, Yukimi and Terumi Sorimachi, for their unlimited kindness. My hearty gratitude goes to my wife, Megumi, and my children, Keita, Tomoki, and Hanaka.

A shorter version of Chapters 4 and 5 appeared in an article in BioSocieties (2022, https://doi.org/10.1057/s41292-022-00283-7). Sections of Chapters 6 and 7 appeared in an article in African Study Monographs (2014, Suppl. 48, 31–47) and in a contribution to Ecologies of Care: Innovations through Technologies, Collectives, and the Senses, edited by Gergely Mohácsi (Osaka University, 2014, pp. 119–31). My research was funded by the JSPS Kakenhi grant numbers 19710210, 20510239, 22510263, 25360011, 18H00785, and Kyoto University GCOE Program: In Search for Sustainable Humanosphere in Asia and Africa.

Contents

1

Introduction

Abstract This chapter concerns the problems of HIV care, which has been profoundly altered by the policies prioritizing the universal use of antiretroviral medicine. The universal treatment represents a process in which humanity's rightful share of this extraordinary fruit of biomedical technology is materialized. However, the process has materialized within a particular tradition of governance that is inherently indifferent to the state of human life. This means that universal treatment became both a condition for human survival and ignorance of the particularities of life and its conditions to prevail. The term "curing lives" represents ideas and actions that resist the increasing ignorance of the shapes of life during pharmaceuticalization. It evokes that life on ART is often full of contingencies and persistent suffering against the prevailing assumption that ART is a "magic bullet" that solves all perceivable problems, whether collective or individual. It calls for ethnographic research to trace the threads of human experience interwoven with particular political, social, and medical arrangements and to ask how people value life and relationships and enact the possibilities they envision for themselves and others.

Keywords Antiretroviral treatment · Experimentation · HIV care · Pharmaceuticalization

HIV intervention in the past decade has been marked by an enormous victory in the universal treatment strategy of making antiretroviral treatment (ART) available to all patients worldwide. Universal treatment is a public health strategy that encompasses two distinct realms of biopolitics: the humanitarian realm that calls for delivering medication to the poor and the health technology realm that calls for the efficient and cost-effective elimination of the pandemic. This book addresses forms of action and engagements through which the suffering inflicted by the epidemic is cared for under the biopolitical conditions induced by the universal treatment strategy. It invites readers to the ambivalent landscape of therapeutic citizenship guided by different voices, including global health experts, policymakers, activists, and community volunteers. Moreover, it focuses on the narratives of two women whose lives are deeply entangled in HIV responses. Both women sought to "cure" the lives of those affected by the epidemic.

To inquire into a life experience under the HIV epidemic is to consider the conditions under which some people are living with excessive suffering while others are allowed, or even encouraged, to remain indifferent. Such conditions include not only the illness stigma (Sontag, 1989) but also intersections of race and sexuality (Woubshet, 2015) and of poverty and violence in the US (DiStefano & Cayetano, 2011), a history of institutionalized racial segregation in South Africa (Fassin, 2007), and the precarious economic and cultural status of women in Haiti (Farmer, 1997) and Kenya (Geissler & Prince, 2010). In the last century, HIV interventions were often designed in a way that associated the disease with moral danger and fear of death, causing people to remain silent on or indifferent toward AIDS (Muthusamy et al., 2009). Such interventions made it difficult for people to talk about the disease (Epstein, 2007) and failed to control the pandemic (Albarracín et al., 2005). It was in such a circumstance that some individuals who were affected by the virus or lost loved ones to AIDS sought a place to share their experiences and grief, making way for varied forms of connect-edness to counter the silence and indifference (Kaleeba & Ray, 2002; Woubshet, 2015). These were the types of solidarity in a world where humans did not have a remedy for AIDS.

However, the basis for such connections was crucially altered by the global dissemination of antiretroviral drugs. Where such treatment became available, the narrative concerning fear and death was replaced with "tales of miraculous resurrection through the power of medicine" (Whyte, 2014a, p. 13). The solidarity of people against silence and indifference found an eminent node—the claim for access to treatment. By the first years of this century, the solidarity for treatment turned into a massive counter-movement against global capitalism. However, at the same time, this was a process in which concern over treatment access replaced that over the suffering of individuals as the focus of the movement.

Pharmaceuticalization of HIV Care

The process has been described as pharmaceuticalization,[1] defined as capturing what happens "when pills are introduced to manage complex health problems," particularly in social settings with limited health and welfare resources (Nguyen, 2015, p. 60). This book addresses how ART rollout has altered the actions and engagements of those affected by HIV (McKay, 2018; Nguyen, 2010; Whyte, 2014b). It also illuminates how the process transformed the modes of HIV intervention and how these embraced or rejected sufferers' voices (Biehl, 2007). In short, pharmaceuticalization profoundly altered the forms of HIV care.

I define care as encompassing a range of engagements and evoking normative claims to address the problems of those affected by the HIV epidemic. Care, on the one hand, is embedded in situated practices (McKay, 2018). Ramah McKay describes care as encapsulating "all the material, social, epistemological, and medical work that accompanies pharmaceuticals" (2018, p. 16). However, care also evokes normative claims. McKay explains that the conceptualizations of care "not only draw attention to health inequities and serve as calls for action; they also reflect normative political claims about what the state is or should be" and "serve to index caring subjects and subjectivities in ways that are raced, classed, and gendered" (p. 16).

In terms of situated engagements and normative claims, the rollout of the free nationwide ART program in Ethiopia profoundly altered the forms of HIV care. I argue that this process has largely been facilitated by "new prevention technologies" (Nguyen, 2015, p. 50), which accompanied the scale-up of HIV treatment in South Africa (Mahajan, 2018), Brazil (Biehl, 2007), and elsewhere in the Global South. Such technologies include treatment-as-prevention (TasP), pre-exposure prophylaxis (PrEP), prevention of mother-to-child transmission (PMTCT), and voluntary male circumcision. Manjari Mahajan argues that these interventions are "resolutely biomedical" (2018, p. 149). In contrast to many older prevention strategies, including behavior change programs and broad-based social measures, new techniques are "animated by logics of calculable efficacy" (Mahajan, 2018, p. 149). It is a great stretch to say that such technologies literary made HIV interventions "calculable" or predictable because, as I will discuss in Chapter 2, those novel forms of intervention unfolded amid much uncertainty. Researchers learned the consequences only after the "experimentations" took place in African societies.

Among these techniques, TasP has played the most significant role in Africa (Nguyen, 2015; Mahajan, 2018), including Ethiopia. The advancement of the TasP strategy has been based on the assumption that the HIV epidemic in a given population may be eliminated by achieving the universal administration of antiretroviral drugs. When properly adhered to, the drugs significantly reduce the risk of viral transmission because they reduce the amount of target virus circulating in bodily fluids (Eisinger et al., 2019). The aggressive expansion of ART to reach all individuals with HIV and upholding medication adherence among those on ART are considered vital for a successful TasP intervention.

Epidemiologists have invested considerable efforts to predict the efficacy of this strategy by constructing mathematical models (Cohen &

Gay, 2010; Granich et al., 2009). These findings were supported by a large-scale trial involving multiple countries (Cohen et al., 2011) and other studies involving human subjects (Smith et al., 2011). Moreover, the outlook that the universal treatment strategy may swiftly and cost-effectively end the HIV epidemic in the Global South has helped consolidate financial assistance from wealthy nations and private funds, facilitating the global expansion of ART during the last decade.

How the scale-up of ART and the accompanying process of pharmaceuticalization altered the landscape of HIV care in Africa has been the subject of medical anthropological inquiries (Nguyen et al., 2011; Moyer, 2015; Kenworthy et al., 2018). Susan Whyte argued, based on her research in Uganda, that life with ART was inherently characterized by "chanciness" or dependence on an uncertain event, occurrence, or relationship (Whyte, 2014a, p. 20). In the city of Kisumu, Kenya, the survival of individuals with HIV often depended on contingent care networks. Ruth Prince argued that the "moral economy of survival" was medicalized by the scaling-up of ART programs by highlighting medication adherence while rendering the realities of life—particularly persistent hunger—invisible (2012, p. 548).

The insecurity of life in Africa has partly been reinforced by the "defunding" of primary health care services under the shadow of ART scale-up (Kenworthy et al., 2018, p. 965). The increasingly technocratic nature of HIV interventions, coupled with austerity policies that dominate international funding agencies, have rendered public health personnel in Mozambique chronically and increasingly overworked and overburdened (Pfeiffer & Chapman, 2015, 2019). Moreover, the existing forms of non-pharmaceutical HIV care are often redefined to fit the technocratic scheme. Since the early stages of the HIV epidemic, home-based care has been practiced in some African countries as a community-level response. When home-based care activity was incorporated into the national HIV program in Mozambique in the early 2000s, the program became more technically comprehensive, but the practice was narrowly redefined to be clinically oriented work (Kalofonos, 2021).

In what follows, this study addresses three interrelated questions. First, it asks how the technocratic scheme, which accompanied the ART scale-up and ultimately rendered the suffering of some people invisible, took place in early twenty-first-century Ethiopia. Second, it explores how some effects of the new HIV prevention and care technology have interplayed with local actions and engagements aiming at surviving the epidemic and "curing the lives" of those sufferers. This question leads to the third and broader inquiry concerning life in Ethiopia. Guided by the narratives of some women I encountered in the course of my ethnographic research, this study asks what it takes to become a woman in contemporary Ethiopia. By carefully contextualizing their engagements with the local realities, it gives shape to their struggles for cures for troubled lives.

Africa as a Site for Experimentations

The TasP strategy to eliminate the HIV pandemic by providing universal access to treatment was an unprecedented public health endeavor in Africa. However, this assemblage of practices soon turned out to be a massive health infrastructure enacted in the continuum of biopolitical experimentations aiming at the production of a healthier population in Africa. Indeed, through the times of imperialism, decolonization, and global health crises, Africa has been a "living laboratory," that is, the site for the production of medical knowledge concerning medicinal plants (Osseo-Asare, 2014), disease control (Tilley, 2011), and clinical trials (Graboyes, 2015). Here I follow Murphy's definition of experimentation as "technical-social assemblies that arrange and gather data about interventions into the world toward the possibility of making something different happen" (Murphy, 2017, p. 80).[2] The TasP strategy turned the HIV intervention in Africa into precisely what Murphy referred to as an epistemic infrastructure or the "assemblages of practices of quantification and intervention conducted by multidisciplinary and multisited experts that became consolidated as extensive arrangements of research and governance" (Murphy, 2017, p. 6).

These are the conditions that defined the shape of therapeutic citizenship in Ethiopian society in the past decade. Therapeutic citizenship is a term coined by medical anthropologist Vinh-Kim Nguyen to refer to "a system of claims and ethical projects that arise out of the conjugation of techniques used to govern populations and manage individual bodies" (Nguyen, 2005, p. 126). The term is built on the idea of biological citizenship, which, according to Nikolas Rose, encompasses two realms of biopolitics. The idea refers to a form of political activism concerning the development and application of biomedical technologies (Rose, 2006, pp. 144–147). However, at the same time, it has to do with certain systems of recognition and exclusion imposed on populations with certain biological features (Rose, 2006, pp. 131–139). As the central architecture for contemporary antiretroviral biopolitics, the TasP strategy has significantly altered the moral premises of therapeutic citizenship in Ethiopia and elsewhere in the Global South. It has rendered the conventional methods, based on fear and moral danger, ineffective and less attractive to policymakers. Furthermore, it has altered how the meanings of the illness and care are shaped. As I elucidate in this book, its effects unfolded in Ethiopia in two seemingly contradictory layers: the one in which silence was broken and solidarity took place, and the other in which care was denied and institutionalized indifference prevailed.

Moreover, this process has resulted in what Murphy described as the "economization of life," or a set of calculative practices that designate human lives as worthy or unworthy of investment (Murphy, 2017, p. 6). In the provincial town where I conducted my fieldwork, such a process was manifested as a localized practice of "triaging" the patients—in a way comparable to how João Biehl (2004) described ART in Brazil—to ensure that the scarcely resourced local institutions for universal treatment continue to function by way of "bypassing" the multiple life burdens borne by some patients. In this sphere of rationality, universal treatment strategy becomes an entry point to the economization of life rather than the basis to claim against it because it is the "rationality through which capitalism finally swallows humanity" (Brown, 2015, as cited in Biehl & Locke, 2017, p. 20).

Curing Lives

Access to ART is an essential condition for all human beings to survive the HIV pandemic.[3] Universal HIV treatment represents a process in which humanity's rightful share of this extraordinary fruit of biomedical technology is materialized. However, the process has materialized within a particular tradition of governance that is inherently indifferent to the state of human life. In short, universal treatment became both a condition for human survival and ignorance of the particularities of life and its conditions to prevail.

In this book, the term "curing lives" represents the ideas and actions that resist the increasing ignorance of the shapes of life during pharmaceuticalization. By evoking this term, I intend to make clear the limited "curing" effects of ART. I am not referring to the inability of current ART to achieve viral eradication (in which HIV is eliminated from the body) or a functional cure (in which HIV does not proliferate and cause harm even when treatment is suspended). Instead, I am arguing against the assumption that ART is a "magic bullet" that solves all perceivable problems, whether collective or individual. ART can cure people of AIDS and prolong their lives. However, life on ART is often full of contingencies and persistent suffering (Whyte, 2014b). The term "curing life" evokes the simple fact that it takes more than ART for an individual to (re)claim their life. I doubt the ignorance of human suffering may be reverted by evoking universal humanitarianism because it is something that emerged after such a claim—the universal right for treatment—had been evoked and pursued.

In my ethnographic work, the term "curing life" is anchored to the experience of two Ethiopian women: in their respective narratives, they make such claims as "*Bizu hiwot adigneallehu*" (I have cured many lives) and "*And-and hiwot adinallech*" (She has cured some lives). These are expressions in Amharic, a language widely used in Ethiopia. Both "*adigneallehu*" and "*adinallech*" are conjugative forms of a verb meaning to save (life from spiritual devastation), to spare (life from biological death), or to cure (someone of disease or illness). The abovementioned expressions might also be translated as "I have *saved* many lives" and "She has *saved* some lives." However, here I chose to use the word *cure* to emphasize a secular focus on the lives and experiences during the HIV epidemic in Ethiopia, as well as to problematize the curing power of antiretroviral medicine. The Amharic term *medhanit*, which means medicine, remedy, or cure, derives from the same verb mentioned above. However, as I discuss in Chapters 4 and 7, the two women claim to have "cured lives" to defend their focus on the particularities of life and its meanings and contingencies in the face of the prevailing shift in focus towards pharmaceutical treatment.

I also intend to situate my work and the idea of curing life in an ethnographic tradition that "acknowledges how power and knowledge form bodies, identities, and meanings, and how inequalities disfigure living while refusing to reduce people to the workings of such forces" (Biehl & Locke, 2017, p. 5). Ethnography in this tradition is a method for exploring the modes of contemporary human experience and subjectivity that intertwine with particular configurations of political, economic, and medical institutions, and ask how, under new conditions, people value life and relationships and enact the possibilities they envision for themselves and others (Biehl et al., 2007, p. 8).

Outline of the Book

In the following chapters, I will trace how the unprecedented experimentation to control the HIV pandemic took its shape at the turn of the century (Chapters 2 and 3) and provide an account of its effects on shaping the subjectivity of a woman who survived the epidemic in Ethiopia (Chapters 4 and 5). Furthermore, I will explore forms of action and engagement that resist the economization of life. I will focus on local struggles that are often embedded in, but not defined by, local moral perspectives, particularly regarding the problems of care and womanhood (Chapters 6 and 7).

Chapter 2 describes the concerns among public health experts that complicated the rollout of ART in Africa. I will focus on an essay that argued against the rollout of ART in Africa by questioning its cost-effectiveness. This essay is a striking example of how arguments in favor of a "fairer allocation of resources" can be appropriated and deployed to deny Africans access to ART. Furthermore, it is revealing of the past and present paradigms of HIV intervention: while it represents some of the diminishing claims made by proponents of behavior-based prevention that dominated HIV intervention in Africa during the 1990s, it also represents contemporary public health experts' growing preoccupation with evidence-based medicine. In the twenty-first century, the obsession with cost-effectiveness has become a key driver of knowledge production in global health.

The rest of the chapter focuses on the debates surrounding "antiretroviral anarchy in Africa." A number of medical scientists, including those who were sympathetic to the fate of Africans with HIV, warned against a rapid rollout of ART in Africa on the grounds that the continent lacked the adequate infrastructure for the proper utilization of antiretrovirals. Some anticipated a grave outcome—a future in which "antiretroviral anarchy" in Africa would create a drug-resistant virus and undermine global efforts to stop the HIV pandemic. However, as the experimentation proceeded, the obsession with infrastructure came to be replaced by an obsession with cost-effectiveness, as will be demonstrated in Chapter 3.

Chapter 3 examines the consequences of ART rollout in Africa. I begin with a sketch of a rural trade center in 1990s Ethiopia with only one health post that was barely maintained. Public health services were hardly a part of rural life in the country during those days. With this sheer absence of health institutions, the Ethiopian government started its endeavor to install a national ART system. Nevertheless, Ethiopia's ART system had emerged as one of the most coordinated ones in sub-Saharan Africa within a decade, thus proving that fears of "antiretroviral anarchy" were exaggerated and the need for an adequate level of infrastructure had been overemphasized.

Epidemiologists quickly interpreted the success of ART in Africa as evidence of the TasP strategy, which continues to dominate HIV interventions worldwide. Drawing on an abundance of data on ART implementation in Africa, they constructed a mathematical model to demonstrate that universal treatment was virtually the only way to eliminate the pandemic cost-effectively. While the finely constructed model provided rationale for further expansion of ART in Africa, it also paved the way for defunding of non-pharmaceutical HIV care, including home-based care activities. The focus on the economic value of intervention and the focus away from the individual state of life is a prominent characteristic of contemporary HIV intervention. I argue that such a tendency is not new to the tradition of health interventions in Africa: TasP strategy is a recent addition to the continuum of biopolitical experimentations aiming at governing the population in the Global South.

The following two chapters discuss some effects of the HIV intervention in shaping subjectivity during the epidemic in Ethiopia. Guided by Meseret's account of her own experiences, I examine how individuals' lives were shaped as the universal treatment strategy unfolded in the country.

Chapter 4 offers an account of the state of life before ART when the very mention of AIDS conjured such feelings of fear that few people even dared to speak of it. After she lost her husband to AIDS and was diagnosed as HIV-positive herself, Meseret was subjected to constant harassment by her in-laws, so much so that she finally decided to leave their village. She sought refuge in Entoto Mountain, where thousands of people "like her" lived in shacks on the mountainside. Meanwhile, connections among those diagnosed as HIV-positive were being forged in Addis Ababa, including one started in the form of a traditional fraternity association.

The time of silence was followed by a moment of extraordinary change when HIV-positive people and organizations rapidly acquired a public presence. This shift was initially facilitated by international HIV activist groups keen to connect with their Ethiopian counterparts. Consequently, a national network of HIV-positive people's associations was in place when the universal ART program started in Ethiopia. Meanwhile, Meseret left Entoto Mountain and moved to a small town in her home district. Her work with local authorities to educate people about HIV caught the attention of key members of the emergent national network. However, Meseret decided to remain in her home district as a leader of a local association. These were formative years for Ethiopia's HIV activists when they departed from where they once were. I suggest that this departure was significant in two ways: it meant a departure from a life lived in fear and silence, and it also meant a departure from close engagements with the suffering of others. Keeping some distance from the national movement allowed Meseret to continue to engage closely with the pain of others like her.

Chapter 5 addresses the realities of life and suffering under the universal ART regime in Ethiopia. It exposes the ambiguous exercise of "triage," that is, the exercise of latently yet systematically distinguishing between those who are "worthy" and those who are "unworthy" of laying claim to a meaningful life. In this regime, local health institutions simultaneously promote universal treatment by disseminating free antiretrovirals and triage by referring those with excessive suffering to Meseret's association. As a result, her group has borne an extraordinary concentration of such individuals. I regard this practice as an exercise in triage because, under the current ART regime in Ethiopia, Meseret's association has been systematically deprived of the resources required to address its members' needs. The emphasis on antiretroviral coverage and cost-effectiveness has resulted in a shrinking focus on social initiatives, such as providing home-based care by local associations like Meseret's. Moreover, the virtual ban on foreign aid to non-governmental organizations in Ethiopia has precluded the possibility of alternative funding. The ART regime in Ethiopia has ensured that those with excessive suffering are rendered invisible in the landscape of therapeutic citizenship in the country.

The last two chapters present an inquiry into culturally informed ideas of care and womanhood, guided by the narratives of two women, Meseret and Asha.

Chapter 6 examines various facets of Meseret's life as embedded in, but not defined by, local moral perspectives, particularly regarding the problems of care and womanhood. In the local social milieu, Meseret's involvement, on the one hand, is often understood in terms of a heroic "fight against AIDS," framing her life as comparable to that of a *jegna* or war hero who protects the community from foreign enemies. On the other hand, Meseret's own account of her life suggests that her moral subjectivity revolves around the image of an *ajyet* or caring woman. To better understand Meseret's moral endeavors in contemporary Ethiopian society, I look into local interpretations and reinterpretations of gender relationships. I also examine local efforts to combat the HIV epidemic, many of which are led by traditional male leaders and driven mainly by the idea of defending rural women from the virus.

Chapter 7 turns to the life and engagements of another Ethiopian woman, Asha, whose life never crossed with that of Meseret. Their personal attributes differ, including their background, religion, and HIV status. Still, I see many similarities between Meseret's work and Asha's continued engagement, as a laywoman, with other people's suffering.

Asha was adopted and raised by a woman who abandoned her family and struggled to make a life in Addis Ababa. Mother and daughter found themselves in a number of detrimental situations and consequent poverty. After her mother's death, Asha found herself in an impoverished settlement where she spent most of her life raising three of her children. Asha's knowledge of midwifery and the leadership roles she assumed in neighborhood groups, such as burial associations, were found useful by the local officials and agents who wished to promote community health. They repeatedly sought her services as a traditional birth attendant or HIV home-visit volunteer, neither of which earned her the remuneration that she deserved. I argue, however, that the seemingly fragmentary pieces of Asha's engagement in her community point to a sociality centered on care and responsiveness that survived patriarchal impositions and differential systems. Moreover, her tireless engagements with neighbors were underpinned by the "culture of defiance" inherited and shared among Ethiopian women.

To shed more light on how Meseret's work is embedded in, but not defined by, local moral perspectives, I present her account of her own childhood. Her aspiration to advance through a normative course toward becoming a "caring woman" was complicated by a series of adverse events and troubled relationships. Her recollection of her mother, who died in isolation, offers a focus for her moral inquiry, which emerges at the intersection of crisis, care, and womanhood.

As Biehl and Locke (2017) put it, the human subject is "not an autonomous, rational individual or a stable self but an always unstable assemblage of organic, social, and structural forces [...] that at once shape and are shaped by their milieus" (p. 8). "Curing lives" is a means of engagement, of participation in the shaping of a moral subjectivity in which one continues to ask how the lives of individuals become meaningful to themselves and those surrounding them within changing social milieus.

Notes on the Author's Research in Ethiopia

I first visited Ethiopia in October 1995 to serve as a staff at the Embassy of Japan in Addis Ababa. I worked for the Embassy for more than five years, from October 1995–December 1997 and August 1999–August 2002, engaging in policy analysis and dealing with small-scale grants to assist non-government organizations. My work experience at the embassy helped me to get familiar with government policies, foreign aid procedures, and the state of civil society activities in the country. I also visited the country repeatedly to conduct ethnographic fieldwork in 1998 and between 2003 and 2015 as a visiting researcher at the Institute of Ethiopian Studies, Addis Ababa University. This book is mainly based on my field research between 2007 (when I first met Meseret) and 2015, though knowledge from my earlier research is also mentioned in some chapters to support my discussion.

Chapters 2 and 3 are primarily based on research articles, government reports, and data provided by funding agencies. The remaining chapters, except for several sections in Chapter 7, are mainly based on my interviews with Meseret Gebre. I use her official name with her consent and based on the fact that she represented an association of people with HIV, making her name and HIV status public. People representing the HIV movement in Ethiopia are also addressed with their official names, while other HIV-positive individuals are addressed with pseudonyms.

My interviews with Meseret took place between September 2007 and October 2015. During our sessions, she shared stories concerning the lives of her own and some fellow HIV-positive Ethiopians, part of which are recounted in this book. Chapters 4, 5, and a section in Chapter 7 are based on my research with Meseret and the members of the Fana association, the group of HIV-positive people she represented. The research included interviews, focus group meetings, and a questionnaire survey. I also attended some of the association's regular meetings. Meseret and I also visited places such as health facilities in and around Welkite town and households in her home village to trace her trajectory. Chapter 6 is mainly based on my interviews and observations in her home village. Some key figures in HIV care and advocacy groups and government health bureaus were also interviewed to support my findings described in this book.

Several sections in Chapter 7 are based on my interviews with Asha Bule, a traditional birth attendant in Addis Ababa. I first met her in 2000 as the mother of one of my research assistants. My early interviews with Asha focused on the activities of burial associations in Addis Ababa, which was one of my research topics before HIV. My conversation with her helped me better to understand the broader aspects of Ethiopian sociality, in which death is a significant focal point. She later shared with me the stories of her life, including her experience as a traditional birth attendant and an HIV volunteer and memories of her mother, who had died in 1973. Our conversation continued until Asha died in 2016.

In my interviews with Meseret, Asha, and other Ethiopian interlocutors, I used Amharic as the medium of communication. Amharic is an Ethiopian Semitic language spoken as lingua franca in the country where more than 80 languages are spoken. Meseret and Asha are fluent speakers of Amharic, although the former is of Gurage origin and the latter is of Oromo.[4] Amharic and Gurage have unique writing systems derived from the Ge'ez script. However, this book uses the italicized Latin alphabet to describe the words and phrases in these languages.[5]

Notes

1. The term has been used broadly in discussions that relate the use of pharmacies with a range of issues, including cultural politics of health control (Whyte et al., 2009), biomedicalism promoted by pharmaceutical innovation (Abraham, 2010), and capitalization of health care (Gaudilliere & Sunder Rajan, 2021).
2. This meaning should not be confused with narrowly defined scientific experiments, such as clinical trials that use randomized control and test

populations. I use the term experimentation to refer to "a more general-
ized and loose mode inclusive of many forms and ideologies" (Murphy,
2017, p. 80) or an ongoing "process of reorientation and reshuffling of
the boundary between what is thought to be known and what is beyond
imagination" (Rheinberger, 1997, p. 5).

3. A notable exception is the rare "elite controllers," or those with the innate
   ability to suppress viral loads (Jiang et al., 2020).
4. Gurage and Oromo are the names of ethnic groups in Ethiopia.
   According to the national census conducted in 2007, the Oromo was
   the largest ethnic group in Ethiopia, consisting of 34.4% of the total
   population, while the Gurage was the sixth with 2.5% (Central Statistical Agency, 2010). Both ethnic groups consist of Christian and Muslim
   people. They have been ruled by the Ethiopian government since the end
   of the nineteenth century when the state territory expanded southward
   (Bahru, 2002).
5. Amharic words are written in this book in a modified version of the
   BGN/PCGN romanization system (UK Government, 2014). All diacrit-
   ical marks in the original system are omitted in my version. Also, *gn*
   replaces *ny* to represent the sound in the " ኝ" family: thus, *adigneallehu*
   instead of *adinyeallehu* (meaning, I have cured). However, some names of
   places and individuals are written following conventions: thus, "Entoto"
   Mountain instead of "Intoto."

References

Abraham, J. (2010). Pharmaceuticalization of society in context: Theoretical,
empirical and health dimensions. *Sociology, 44*(4), 603–622.

Albarracín, D., et al. (2005). A test of major assumptions about behavior
change: A comprehensive look at the effects of passive and active HIV-
prevention interventions since the beginning of the epidemic. *Psychological
Bulletin, 131*(6), 856–897.

Bahru, Z. (2002). *A history of modern Ethiopia, 1855–1991* (2nd ed.). Addis
Ababa University Press.

Biehl, J., & Locke, P. (2017). Introduction: Ethnographic sensorium. In J. Biehl & P. Locke (Eds.), *Unfinished: The anthropology of becoming* (pp. 1–38). Duke University Press.

Biehl, J. (2004). The activist state: Global pharmaceuticals, AIDS, and citizenship in Brazil. *Social Text, 22*(3), 105–132.

Biehl, J. (2007). Pharmaceuticalization: AIDS treatment and global health politics. *Anthropological Quarterly, 80*(4), 1083–1126.

Brown, W. (2015). *Undoing the demos: Neoliberalism's stealth revolution*. Zone Books.

Central Statistical Agency. (2010). *Population and housing census of 2007: Report for southern nations, nationalities and peoples' region, part 1: Population size and characteristics*. Central Statistical Agency, Federal Democratic Republic of Ethiopia.

Cohen, M. S., & Gay, C. L. (2010). Treatment to prevent transmission of HIV-1. *Clinical Infectious Diseases, 50*(Suppl 3), S85–S95.

Cohen, M. S., et al. (2011). Prevention of HIV-1 infection with early antiretroviral therapy. *The New England Journal of Medicine, 365*(6), 493–505.

DiStefano, A. S., & Cayetano, R. T. (2011). Health care and social service providers' observations on the intersection of HIV/AIDS and violence among their clients and patients. *Qualitative Health Research, 21*(7), 884–899.

Eisinger, R. W., Dieffenbach, C. W., & Fauci, A. S. (2019). HIV viral load and transmissibility of HIV infection: Undetectable equals untransmittable. *JAMA, 321*(5), 451–452.

Epstein, H. (2007). *The invisible cure: Africa, the West, and the fight against AIDS*. Picador.

Farmer, P. (1997). On suffering and structural violence: A view from below. In A. Kleinman, V. Das, & M. Lock (Eds.), *Social suffering* (pp. 261–283). University of California Press.

Fassin, D. (2007). *When bodies remember: Experiences and politics of AIDS in South Africa*. University of California Press.

Gaudilliere, J.-P., & Sunder Rajan, K. (2021). Making valuable health: Pharmaceuticals, global capital and alternative political economies. *BioSocieties, 16*(3), 313–322.

Geissler, P. W., & Prince, R. J. (2010). *The land is dying: Contingency, creativity and conflict in western Kenya*. Berghahn Books.

Graboyes, M. (2015). *The experiment must continue: Medical research and ethics in East Africa, 1940–2014*. Ohio University Press.

Granich, R. M., et al. (2009). Universal voluntary HIV testing with immediate antiretroviral therapy as a strategy for elimination of HIV transmission: A mathematical model. *The Lancet, 373*(9657), 48–57.

Jiang, C., et al. (2020). Distinct viral reservoirs in individuals with spontaneous control of HIV-1. *Nature, 585*(7824), 261–267.

Kaleeba, N., & Ray, S. (2002). *We miss you all.* SAfAIDS.

Kalofonos, I. (2021). *All I eat is medicine: Going hungry in Mozambique's AIDS economy.* University of California Press.

Kenworthy, N., Thomann, M., & Parker, R. (2018). From a global crisis to the "end of AIDS": New epidemics of signification. *Global Public Health, 13*(8), 960–971.

Mahajan, M. (2018). Rethinking prevention: Shifting conceptualizations of evidence and intervention in South Africa's AIDS epidemic. *BioSocieties, 13*(1), 148–169.

McKay, R. (2018). *Medicine in the meantime : The work of care in Mozambique.* Duke University Press.

Moyer, E. (2015). The Anthropology of life after AIDS: Epistemological continuities in the age of antiretroviral treatment. *Annual Review of Anthropology, 44*(1), 259–275.

Murphy, M. (2017). *The economization of life.* Duke University Press.

Muthusamy, N., Levine, T. R., & Weber, R. (2009). Scaring the already scared: Some problems with HIV/AIDS fear appeals in Namibia. *Journal of Communication, 59*(2), 317–344.

Nguyen, V.-K. (2005). Antiretroviral globalism, biopolitics, and therapeutic citizenship. In A. Ong & S. J. Collier (Eds.), *Global assemblages: Technology, politic, and ethics as anthropological problems* (pp. 124–144). Blackwell.

Nguyen, V.-K. (2010). *The republic of therapy: Triage and sovereignty in West Africa's time of AIDS.* Duke University Press.

Nguyen, V.-K. et al. (2011). Remedicalizing an epidemic: From HIV treatment as prevention to HIV treatment is prevention. *AIDS* 25(3), 291–293.

Nguyen, V.-K. (2015). Treating to prevent HIV: Population trials and experimental societies. In P. W. Geissler (Ed.), *Para-states and medical science: Making African global health* (pp. 47–77). Duke University Press.

Osseo-Asare, A. D. (2014). *Bitter roots: The search for healing plants in Africa.* University of Chicago Press.

Pfeiffer, J., & Chapman, R. (2015). An anthropology of aid in Africa. *The Lancet, 385*(9983), 2144–2145.

Pfeiffer, J., & Chapman, R. R. (2019). NGOs, austerity, and universal health coverage in Mozambique. *Globalization and Health, 15*(Suppl 1), 0.

Prince, R. (2012). HIV and the moral economy of survival in an East African City. *Medical Anthropology Quarterly, 26*(4), 534–556.

Rheinberger, H.-J. (1997). *Toward a history of epistemic things: Synthesizing proteins in the test tube*. Stanford University Press.

Rose, N. (2006). *The politics of life itself*. Princeton University Press.

Smith, K., et al. (2011). HIV-1 treatment as prevention: The good, the bad, and the challenges. *Current Opinion in HIV and AIDS, 6*(4), 315–325.

Sontag, S. (1989). *AIDS and its metaphors*. Farrar, Straus and Giroux.

Tilley, H. (2011). *Africa as a living laboratory: Empire, development, and the problem of scientific knowledge, 1870–1950*. University of Chicago Press.

UK Government. (2014). *Guidance on the US Board on Geographic Names (BGN)/Permanent Committee on Geographical Names (PCGN) romanization systems*. https://www.gov.uk/government/publications/romanization-systems. Accessed January 11, 2023.

Whyte, S., van der Geest, S., & Hardon, A. (Eds.). (2009). *Social lives of medicines*. Cambridge University Press.

Whyte, S. R. (2014a). Introduction: The first generation. In S. R. Whyte (Ed.), *Second chances: Surviving AIDS in Uganda* (pp. 1–24). Duke University Press.

Whyte, S. R. (Ed.). (2014b). *Second chances: Surviving AIDS in Uganda*. Duke University Press.

Woubshet, D. (2015). *The calendar of loss: Race, sexuality, and mourning in the early era of AIDS*. Johns Hopkins University Press.

2

Initiating a New Experimentation

Abstract This chapter describes the concerns among public health experts that complicated the rollout of ART in Africa. It focuses on an essay that argued against the rollout of ART in Africa by questioning its cost-effectiveness. This essay is a striking example of how arguments favoring a "fairer allocation of resources" can be appropriated and deployed to deny Africans access to ART. It also concerns the debates surrounding "antiretroviral anarchy in Africa" at the turn of the last century. A number of medical scientists, including those who were sympathetic to the fate of Africans with HIV, warned against a rapid rollout of ART in Africa on the grounds that the continent lacked the adequate infrastructure for the proper utilization of antiretrovirals. Some anticipated a grave outcome—a future in which "antiretroviral anarchy" in Africa would create a drug-resistant virus and undermine global efforts to stop the HIV pandemic. However, as the experimentation proceeded, the obsession with infrastructure came to be replaced by an obsession with cost-effectiveness, as will be demonstrated in Chapter 3.

Keywords Antiretroviral anarchy · Behavior-based prevention · Health infrastructure · Moral approach

Now, for perhaps the first time in history, we must decide whether economic reality will permit an informed debate about rationing that could result in millions of patients receiving supportive care, but not treatment, to prevent many more millions from becoming afflicted with the disease. The allocation of resources, the development of a just health-care system, and the adjudication of the rights and claims of competing groups [...] are the present moral issues facing HIV policy makers in sub-Saharan Africa. (Marseille et al., 2002, p. 1855)

Public health experts Elliot Marseille and colleagues made the above statement in 2002 as the large-scale expansion of antiretroviral treatment in Africa was about to take place. During this period, the HIV activists' demand for universal treatment forced pharmaceutical companies to reduce drug prices in low-income countries. Marseille and colleagues claimed preventive intervention was 28 times more cost-effective than ART. They argued that withholding ART to allocate more resources for prevention was a "required" option, even though the choice meant, as they indirectly suggested, that millions of Africans would continue to die of AIDS. They further argued that those who demanded ART for Africans were essentially appealing for the "rule of rescue," a form of humanitarian reasoning directed toward spending more resources for "identified sufferers" (who were dying of AIDS) than for "future sufferers" (who would get HIV). They were convinced that the rule of rescue was problematic because "it does not consider the wisdom of their application in economic or comparative cost-effectiveness terms" (Marseille et al., 2002, p. 1855).

HIV intervention in Africa after 2002 turned out to be something vastly different from that envisaged by the proponents of behavior-based preventive interventions, as I will make clear. However, it is worth examining first how these proponents articulated their ideas by leaping across different realities—between North America and sub-Saharan Africa. Marseille and colleagues drew their idea from a short essay titled "Bentham in a box: Technology assessment and health care allocation," written by medical ethicist Albert Jonsen (1986). In this essay, Jonsen questioned the continued investment in health technologies in a way that benefited "a few at cost to many." More precisely, he was questioning Americans' motivation in investing in such expensive technology as heart transplantation and renal dialysis when they could save more lives by investing the same amount of money in preventive public health measures. He answered the question by arguing that "technology assessment" or the "rational effort to evaluate the efficacy and costs, the burdens and benefits, of the panoply of medical technologies" was always limited by the "rule of rescue," the deontological imperative to rescue the visible individual from imminent death (Jonsen, 1986, p. 174).

To illuminate his idea, Jonsen suggested that the state of technology assessment in the US resembled that of Jeremy Bentham's auto-icon (a mummified figure) placed in a glass box and displayed at the University College London. For Jonsen, Bentham in a box symbolized "the lifeless, impotent relic of a powerful and vital way of thinking about, and dealing with, the world." Undertaking technology assessment in modern medicine was akin to becoming a Bentham (or performing a rational calculation of human welfare) firmly confined in a box (or limited by the rule of rescue). Although he explained his comparison as "a somewhat facetious allusion," he also suggested that technology assessment in the mid-1980s remained, "in essence, a copy of Bentham's proposal to plan and effect a rational course of action and to create a rational world." He further clarified that while health policy planners were "not utilitarians in theory," they were, "to some extent, utilitarians in practice" (Jonsen, 1986, pp. 172–173).

Jonsen's essay was meant to inform discussions concerning healthcare resource allocation in the US when the Regan administration increased its attack on the federal subvention of health care. Jonsen wrote the essay in response to the report produced by the President's Commission for the Study of Ethical Problems in Medicine and Biomedical Research (1983). This report was written in defense of the "ultimate responsibility" of the federal government to arrange for equitable access to health. However, Jonsen found its definition of "adequate care" too abstract and having little utility to resolve the practical problems concerning just and fair healthcare allocation.[1] Thus, he argued that "one way to descend from that abstractness is to inspect with great attention the various forms of health and medical care, in order to determine the extent to which they actually do effect, on the whole, the benefits desired" (Jonsen, 1986, p. 173).

It is striking to learn how the argument, intended to direct more attention to "invisible multitudes who may die of exposure to toxic chemicals, cigarette smoke, or radiation, or those deprived of immunization or adequate nutrition" in the US (Jonsen, 1986, p. 174), was appropriated in the justification for withholding the use of ART in Africa. Such an appropriation seemed to stand only when one framed the problem as that of resource allocation between prevention and treatment and refused to consider the then-prevailing reality that Africans were denied access to antiretroviral drugs while those in richer countries were taking them.

Looking back from where we are now, we may safely conclude that Marseille and colleagues' essay represents one of the diminishing claims made by proponents of behavior-based prevention that dominated HIV intervention in Africa during the 1990s. The world we know today is different from what was perceived by those who engaged in the "prevention versus treatment" argument in 2002: approximately 29 million people had access to ART worldwide in 2021, 20 million of whom lived in sub-Saharan Africa.[2] Today, few infectious disease epidemiologists agree that behavior-based prevention can be more cost-effective than ART as the means to control the HIV epidemic.

Nonetheless, the future envisaged by Marseille and colleagues prevailed in some twisted way. Since 2002, Africa has become a site for "experiment" where the input and outcomes of investment in health are examined and debated, allowing policymakers to make informed choices over the health of millions. Health policy planners are poised to use their knowledge "to resolve the practical problems of just and fair allocation" in the way Jonsen would have wished to see. It is as if Bentham has been set free from the box in London and is roaming across the African continent. The way it happened was not straightforward in any sense, however.

Tripping into Anarchy

The July 2002 issue of *Nature Medicine* carried a commentary written by epidemiologists at the Imperial College of Science, Technology and Medicine, London, titled, "Antiretroviral therapy to treat and prevent HIV/AIDS in resource-poor settings." The essay starts as follows:

> Prior to the XIIIth World AIDS Conference in Durban, debate raged in the community over the practicalities, cost and ethics of delivering antiretroviral drugs to the developing world. Two years on, that discussion is history. Political and financial forces will now deliver these drugs to such nations. How then can we expect their arrival to alter the HIV landscape? (Garnett et al., 2002, p. 651)

The conference in Durban held in July 2000 was an epochal moment in the history of global health. "Breaking the silence" was the theme of this conference, which was attended by 12,437 delegates from 180 countries. This was when the global HIV activism "finally and forcefully" succeeded in focusing the world's attention on the epidemic in Africa and the developing world (Friedland, 2000). Nine months later, Kofi Annan, then UN Secretary-General, called for the creation of a global fund to help poor countries with their AIDS programs, including the provision of ART (Salim & Karim, 2001). The Global Fund to Fight AIDS, Tuberculosis and Malaria (hereafter, the Global Fund) started operation in January 2002. As global HIV activism was rapidly succeeding in materializing ART access among the poor, researchers in public health and infectious disease epidemiology were anxiously observing the changing landscape of HIV intervention in the developing world. Some took clear positions, whereas others were uncertain of the consequences of the ART expansion in Africa.

On one side, there were Gavin Yamey, deputy editor of the *Western Journal of Medicine*, and William Rankin, co-founder of the Global AIDS Interfaith Alliance, who decisively joined the activists' voice. Their essay appeared in the January 26th issue of the *British Medical Journal*, titled "AIDS and Global Justice." They demanded, citing John Rawls'

Theory of Justice (Rawls, 1999), the distribution of antiretroviral drugs to the world's poorest people (Yamey & Rankin, 2002).

On the other side, psychologists Danielle Popp and Jeffrey D. Fisher (2002) envisaged a grave future, warning: "It is entirely unclear what effect these efforts will have on the many millions of people in developing countries already infected with HIV" (p. 676). Their concern revolved around the problem of medication adherence. They demanded that ART be adhered to correctly and consistently because otherwise, "individual patients may not benefit, may become treatment resistant, and developing countries could become a veritable 'petri dish' for new, treatment-resistant HIV strains" (p. 677). This meant that "the seemingly humanitarian efforts of drug companies, governments, and the UN" might result in "explosive unintended negative consequences." They also noted that "ART regimens can be very complex, and adherence to them is difficult even under optimal conditions, e.g. a reliable supply of medication, an effective healthcare infrastructure, and adequate food, clean water, and electricity." It was clear that "many of these conditions do not exist in developing countries" (p. 677).

Researchers knew that, if adhered to properly, ART would save millions of Africans. However, they were also aware that the price of failure would be enormous. Unfortunately, it was the time when so-called Africa pessimism was at its height among international research and development circles. The 1990s saw increasing reports of violent conflicts across the continent, and political scientists asserted that African states were not only unstable but that a number of them were failing (Herbst, 1996; Mazrui, 1995). It was in 2000 that the grave situation led the *Economist* to name Africa the "hopeless continent," and the World Bank questioned if the continent could "claim the twenty-first century" (World Bank, 2000).

Where the state was not failing, leadership was untrustworthy. In South Africa, which had (and still has) the largest HIV-positive population in the world, HIV denialists persuaded President Thabo Mbeki to state that AIDS was not caused by HIV but was instead a result of poverty. Where leadership seemed strong, people were not responding. In Botswana, HIV prevalence among adults exceeded 25% in 1999 despite the nationwide campaign organized by one of Africa's richest and most stable governments. Uganda was successful at reducing the number of new infections through the 1990s, but one could still wonder if Uganda's success could be replicated in different contexts (Low-Beer & Stoneburner, 2003) or to what extent the encouraging reports reflected the reality on the ground (Parkhurst, 2002).

In those days, promoting ART coverage in Africa was perceived as similar to "tripping into anarchy" (Horton, 2000), even in the eyes of medical scientists who were sympathetic to Africans with HIV. Particularly horrifying was a story from Harare, Zimbabwe. Pharmacologist Nyazema and colleagues observed physicians prescribing and pharmacists stocking antiretroviral drugs in Harare, noting what "appeared to be therapeutic anarchy in the private sector in Harare in the way [antiretrovirals] were being used." According to their report, "prescribers and dispensers were utilising any [antiretrovirals] that they could lay their hands on. It appeared that no effort had been made to develop treatment guidelines that could be followed" (Nyazema et al., 2000, as cited in Horton, 2000, p. 1542).

This "stark failure of infrastructure" led Richard Horton, editor-in-chief of the *Lancet*, to become cautious about the optimistic outlook provided by the advocates of universal treatment. A report prepared in 2000 by the Voluntary Service Overseas (one of the international nonprofit organizations that joined the ART advocacy) stated, "Greater access to essential medicines would make an immediate impact on the lives of those people most affected by HIV/AIDS." Citing this statement, Horton (2000) commented: "This diagnosis neglects the almost insuperable difficulties facing healthcare systems that can hardly begin to administer drug regimens safely and effectively even if those drugs were freely available" (p. 1542). While recognizing the need to make low-cost generic antiretroviral drugs available to lower-income countries, Horton concluded: "Drug provision alone is not the simple solution for patients with HIV/AIDS that some critics suggest. An equal, although less politically galvanising, concern must remain the creation of sustainable primary care services combined with clear guidelines for treatment. Without this basic framework, the provision of antiretroviral drugs will produce chaos, not control" (p. 1542).

It was thus "not just a matter of providing antiretroviral drugs, but also that they must be provided within a structured framework" (Harries et al., 2001, p. 410). Health economist Warren Stevens and colleagues (2004) also predicted that in Africa, a higher proportion of patients were "likely to fall into the category of potential poor adherers" unless resource-intensive adherence programs were made available (p. 281). The problem was that no one was sure exactly what it would take to ensure medication adherence and prevent the replication and spread of a drug-resistant virus. In high-income countries, it was already established that therapy should be routinely monitored by assaying plasma viral loads, which helped clinicians find signs of a drug-resistant virus that may emerge in a patient's body. However, the necessary laboratory infrastructure for monitoring was unavailable in most parts of Africa.

Harries and colleagues (2001), who were involved in tuberculosis control in Malawi, suggested that the framework for ART administration in Africa should replicate the tuberculosis program already in place in many African countries. The key strategy of the program is referred to as Directly Observed Treatment, Short course, or DOTS and involves close monitoring of all patients under therapy. However, as Stevens and colleagues (2004) pointed out, DOTS itself had met with mixed success: treatment completion rates reported from African countries ranged from low (37%) to moderate (78%). Furthermore, there is an important difference between tuberculosis and HIV treatments: while the standard tuberculosis treatment under DOTS lasts from six to nine months, ART is lifelong. A lifelong version of DOTS would involve unforeseen technical difficulties, and the financial burden to sustain it for decades seemed unpredictable. Garnett and colleagues (2002) further argued that "the risks of the emergence and spread of drug resistance in HIV are much greater than with tuberculosis, owing to the high replication rate of the virus and its rapid evolution under the intense selection imposed by drug therapy" (p. 651). Citing the reports from the Ivory Coast (Adjé et al., 2001) and Gabon (Vergne et al., 2002), they further warned that high levels of drug resistance were already reported from countries where access to treatment was patchy.

Thus, Garnett and colleagues (2002) found it "difficult to envisage the successful use of antiretrovirals to control the spread of infection"

(p. 653). However, at the same time, they were convinced that the "emergence of drug-resistant viral phenotypes in resource-poor regions should not be viewed as grounds for limiting access to drugs." They believed that even imperfect treatment would have "both direct benefits to the patient (reduced morbidity and mortality) and indirect benefits to the community (reduced transmission)." They indicated that it was "reasonable to assume that reducing plasma viral load by treatment will counteract any increase in incidence associated with the longer survival (and potential infectiousness) of those treated" (p. 653). If their position toward ART expansion in Africa—talking about its benefits while being skeptical about its success—seems inconsistent, it is none of their fault. They knew that the global expansion of ART could not be stopped anyway and that they would be dragged into the perceived "anarchy" of the African continent.

Altering Institutional Arrangements

In his memoir, Peter Piot (2012), a Belgian microbiologist and the first Executive Director of the Joint United Nations Programme on HIV/AIDS (UNAIDS), recalls the state of treatment expansion by the year 2001: "The launch of UNAIDS in 1996 coincided with the discovery of antiretroviral treatment, but five years later we had made little progress." His effort to expand ART access in Africa had been met with "long lists of why HIV treatment was not feasible" (p. 299). In their correspondence to the *Lancet*, Piot and colleagues (2002) refuted the claim of Marseille and colleagues (2002) that prevention was more cost-effective than treatment. The former asserted that the latter fundamentally misstated the problem by overlooking the devastating effect of AIDS on Africa's public health systems and national economies. They further argued that the cost of treatment would not stay the same: "The cost ratio of care to prevention may currently be 28 to 1. 2 years ago it was more than 200 to 1. In 2 years time it will probably be far lower than it is today" (Piot et al., 2002, p. 86).

However, it would be misleading to label Piot as a "pro-treatment" advocate in the treatment versus prevention debate at the turn of the century. He had been open to any strategy that would curve the massive death toll from AIDS. While vigorously promoting ART expansion in Africa, he was one of the co-authors of an article that called for a global HIV vaccine enterprise (Klausner et al., 2003). Despite many trials of possible HIV vaccines, their safety and efficacy as preventive options have not been established to date. In 2008, he was the first author of an article that called for more prevention measures (Piot et al., 2008).

Piot's focus on practical gains—mobilizing large amounts of money, in particular—led him to negotiate with people outside the medical circle. Through his experience as the leader of UNAIDS, Piot learned that "doing a reasonable job in terms of epidemiology and of formulating technical solutions" was not enough to have a voice in the "world of political power, where big decisions are made." By 2001, he had reached the conclusion that "in international politics there are only two things that count: the economy and security. As they say in France, the rest is just literature" (Piot, 2012, pp. 247–248).

Indeed, in the language of economy, he convinced leaders of the political and business world to invest on a vast scale in the fight against AIDS. Yet, to pursue his project, Piot had to face another enemy: the institutional legacy of twentieth-century international health. As Piot (2012) notes in his memoir: "Where did the resistance stem from? Not (yet) from the virus, but from institutions and experts" (p. 299). He was so frustrated by the fact that, as late as 1998, the World Health Organization (WHO) "explicitly did not want to be associated with our efforts. The experts at WHO's Program on Essential Drugs thought it was foolish to introduce high-tech medication in developing countries when they were struggling to distribute basic drugs for malaria" (p. 306). HIV activists, too, were pointing at WHO as a chief enemy—after the international pharmaceuticals that decisively resisted the idea of providing poor nations with cheaper generic antiretrovirals at that time. On July 11, 2000, when the AIDS Conference in Durban was underway, the HIV activist group ACT UP harshly criticized WHO's vague attitude toward treatment access:

Only two days ago, after UNAIDS showed data proving that only competition between generic and brand-name products has ever lead to substantial drug price reductions, Mr Tarantola, political advisor to Mrs Brundtland, Director of WHO, made it his business to provide a diversion: "price is but the least obstacle." He went on to list all the supposed prerequisites to treatment access: clean water, hygiene, nutrition, human rights, etc. When will WHO outgrow its caricature of a vision of developing countries, and cease to see them as endless barren lands crowded with starving throngs soaking in backward ignorance? (ACT UP, 2000)

One of the issues that complicated WHO's HIV intervention policy was the Essential Drugs List (now renamed the Essential Medicines List), one of the central tenets of WHO's international health strategy since the 1970s. WHO introduced the idea as part of its effort to help developing countries improve primary health care coverage while limiting the cost of drug procurement (World Health Organization, 1977). The salient feature of the organization's definition of essential drugs was that it preferred "older drugs of proven efficacy and safety over newer drugs" and "generic names over brand names" (Greene, 2011, p. 18). While the idea of essential drugs caused considerable discontentment among international pharmaceuticals, it was welcomed by postcolonial nations. As a result of explosive pharmaceutical developments in the post-WWII era, the market in the Global South was overwhelmed with an array of products of varied quality and novelty. This caused considerable regulatory and economic burdens on developing countries. By the early 1970s, Brazil and Argentina, respectively, had 24,000 and 17,000 brand-name drugs, whereas Norway allowed a market of only 1000 drug brands. In 1976, Thailand spent 30.4% of its public health budget on drugs, whereas Bangladesh spent 63.7% of its budget on prescription medicines (Greene, 2011, p. 16). Postcolonial nations, which were gaining a voice within the UN system, saw the idea of essential drugs as a protection against the profit-hungry multinational pharmaceutical corporations based in the North.

However, the historical context surrounding the idea of essential drugs had largely been altered by the 1990s, when proponents of HIV treatment access started to perceive the idea of essential drugs as a serious impediment to their goal. In March 2001, a debate was organized by the

Royal Society of Tropical Medicine and Hygiene to discuss the plausibility of the list that had been increasingly condemned as "a real barrier to the access of developing countries to modern forms of therapy" (Breckenridge, 2003, p. 1). At this debate, Pierre Chirac, representing Médecins Sans Frontières (MSF), expressed his organization's concern that the fact that WHO excluded patented drugs from the list undermined the efforts of some Global South governments to make newer drugs more accessible to its citizens. He went on to point out that "a disease that kills three million people every year, AIDS, is excluded from the list" (Chirac, 2003, pp. 10–11).

It was in June 2002, six months after the Global Fund was formally established to provide money for low- and middle-income countries to procure and distribute antiretrovirals, that WHO finally published its first guidelines for the use of antiretrovirals in resource-poor settings and expressed its commitment to the global alliance for expanding access to ART (World Health Organization, 2002). However, the organization's position in the political and institutional settings of global health had largely been altered by that time. Medical historian Jeremy Greene precisely summed up the shift in the institutional landscape by tracing the essential drugs debate: "The re-emergence of essential medicines as a critical discourse in the late 1990s and 2000s would largely be driven by forces outside of the WHO, which placed pharmaceuticals at the center of a headless assemblage of actors and institutions all claiming to work in the new field of 'global health.' In place of a central monolithic bureaucracy of world health, the new field of global health involves the competition of multiple overlapping private, public and public–private hybrid institutions within a marketplace of global health delivery" (Greene, 2011, p. 22).

To sum up, this chapter highlighted two issues that complicated HIV treatment access in Africa and elsewhere in the Global South: the perceived "anarchy" in the African continent and the institutional legacy of twentieth-century international health.[3] These issues are strongly associated with the key features of universal HIV treatment as the pioneering global health experimentation. Experts' fear of antiretroviral anarchy reflected their obsession with adequate infrastructure, that is, a combination of well-trained health personnel and well-equipped health facilities that covered entire populations, coupled with well-developed guidelines and regulations to ensure proper use of antiretrovirals, all of which were absent in Africa. However, as I will demonstrate in the following chapter, the obsession with infrastructure has been replaced by that with cost-effectiveness during the global rollout of ART. Furthermore, altering the institutional arrangements of twentieth-century international health was a necessary condition for this transition. The presence of a single dominant bureaucracy that claims to regulate the entire process was increasingly perceived as inhibiting the swift and cost-effective operation to fight against the HIV pandemic. The new arrangements featured multiple actors, governments, private funds, North- and South-based pharmaceuticals, NGOs, and associations of HIV-positive people, with several international institutions such as WHO, UNAIDS, and the Global Fund serving as hubs of the global network.

Notes

1. The President's Commission report concluded that it was the duty of society to assure equitable access to an adequate level of care for all. It described an adequate level as enough care to achieve sufficient welfare, opportunity, information, and evidence of interpersonal concern to facilitate a reasonably full and satisfying life (Jonsen, 1986, p. 173).
2. Data extracted from UNAIDS AIDSinfo, http://aidsinfo.unaids.org. Accessed on October 28, 2022. Unless otherwise specified, other epidemiological figures in this work refer to the same database.
3. Obviously, there is at least one more serious issue related to my discussion here: intellectual property rights as the major barrier to producing and distributing cheap generic antiretrovirals for patients in the Global South. See Heimer (2007) for an extensive review of the literature concerning the issue of intellectual property, pricing, and access to antiretroviral drugs in sub-Saharan Africa.

References

ACT UP. (2000). W.H.O. sold out to big pharma. *ACT UP Historical Archive.* https://actupny.org/reports/durban-who.html. Accessed on March 13, 2021.

Adjé, C., et al. (2001). High prevalence of genotypic and phenotypic HIV-1 drug-resistant strains among patients receiving antiretroviral therapy in Abidjan, Cote d'Ivoire. *Journal of Acquired Immune Deficiency Syndromes, 26*(5), 501–506.

Breckenridge, A. (2003). Royal Society of Tropical Medicine and Hygiene Meeting at the University of Liverpool, Liverpool, 16 March 2001 Debate that "This house believes the essential drug concept hinders the effective deployment of drugs in developing countries." *Transactions of the Royal Society of Tropical Medicine and Hygiene, 97*(1), 1.

Chirac, P. (2003). Translating the essential drugs concept into the context of the year 2000. *Transactions of the Royal Society of Tropical Medicine and Hygiene, 97*(1), 10–12.

Friedland, G. H. (2000). Breaking the silence. *AIDS Clinical Care, 12*(8), 63–69.

Garnett, G. P., et al. (2002). Antiretroviral therapy to treat and prevent HIV/AIDS in resource-poor settings. *Nature Medicine, 8*(7), 651–654.

Greene, J. A. (2011). Making medicines essential: The emergent centrality of pharmaceuticals in global health. *BioSocieties, 6*(1), 10–33.

Harries, A. D., et al. (2001). Preventing antiretroviral anarchy in sub-Saharan Africa. *The Lancet, 358*(9279), 410–414.

Heimer, C. A. (2007). Old inequalities, new disease: HIV/AIDS in sub-Saharan Africa. *Annual Review of Sociology, 33*(1), 551–577.

Herbst, J. (1996). Responding to state failure in Africa. *International Security, 21*(3), 120–144.

Horton, R. (2000). African AIDS beyond Mbeki: Tripping into anarchy. *The Lancet, 356*(9241), 1541–1542.

Jonsen, A. R. (1986). Bentham in a box: Technology assessment and health care allocation. *The Journal of Law, Medicine & Ethics, 14*(3–4), 172–174.

Klausner, R. D., et al. (2003). Enhanced: The need for a global HIV vaccine enterprise. *Science, 300*(5628), 2036–2039.

Low-Beer, D., & Stoneburner, R. L. (2003). Behaviour and communication change in reducing HIV: Is Uganda unique? *African Journal of AIDS Research, 2*(1), 9–21.

Marseille, E., Hofmann, P. B., & Kahn, J. G. (2002). HIV prevention before HAART in sub-Saharan Africa. *The Lancet, 359*(9320), 1851–1856.

Mazrui, A. A. (1995). The blood of experience: The failed state and political collapse in Africa. *World Policy Journal, 12*(1), 28–34.

Nyazema, N. Z., et al. (2000). Antiretrovial (ARV) drug utilisation in Harare. *The Central African Journal of Medicine, 46*(4), 89–93.

Parkhurst, J. O. (2002). The Ugandan success story? Evidence and claims of HIV-1 prevention. *The Lancet, 360*(9326), 78–80.

Piot, P. (2012). *No time to lose: A life in pursuit of deadly viruses.* W. W. Norton.

Piot, P., et al. (2008). Coming to terms with complexity: A call to action for HIV prevention. *The Lancet, 372*(9641), 845–859.

Piot, P., Zewdie, D., & Türmen, T. (2002). HIV/AIDS prevention and treatment. *The Lancet, 360*(9326), 86.

Popp, D., & Fisher, J. D. (2002). First, do no harm: A call for emphasizing adherence and HIV prevention interventions in active antiretroviral therapy programs in the developing world. *AIDS, 16*(4), 676–678.

President's Commission for the Study of Ethical Problems in Medicine and Biomedical and Behavioral Research. (1983). *Securing access to health care: A report on the ethical implications of differences in availability of health services, Volume one: Report.* President's Commission for the Study of Ethical Problems in Medicine and Biomedical and Behavioral Research.

Rawls, J. (1999). *A theory of justice* (Revised). Belknap Press of Harvard University Press.

Salim, S. A. K., & Karim, Q. A. (2001). Breaking the silence, one year later: Reflections on the Durban conference. *AIDS Clinical Care, 13*(7), 63–65.

Stevens, W., Kaye, S., & Corrah, T. (2004). Antiretroviral therapy in Africa. *British Medical Journal, 328*(7434), 280–282.

Vergne, L., et al. (2002). Resistance to antiretroviral treatment in Gabon: Need for implementation of guidelines on antiretroviral therapy use and HIV-1 drug resistance monitoring in developing countries. *Journal of Acquired Immune Deficiency Syndromes, 29*(2), 165–168.

World Bank. (2000). *Can Africa claim the 21st century?* The World Bank.

World Health Organization. (1977). *The selection of essential drugs* (WHO Technical Report Series no. 614). World Health Organization.

World Health Organization. (2002). *Scaling up antiretroviral therapy in resource-limited settings.* World Health Organization.

Yamey, G., & Rankin, W. W. (2002). AIDS and global justice. *British Medical Journal, 324*(7331), 181–182.

3

Installation of a Health System

Abstract This chapter examines the consequences of ART rollout in Africa. Ethiopia's ART system emerged as one of the most coordinated in sub-Saharan Africa during the first decade of the twenty-first century, thus proving that fears of "antiretroviral anarchy" were exaggerated and the need for an adequate level of infrastructure had been overemphasized. Epidemiologists quickly interpreted the success of ART in Africa as evidence of the treatment-as-prevention (TasP) strategy, which continues to dominate HIV interventions worldwide. Drawing on abundant data on ART implementation in Africa, they constructed a mathematical model to demonstrate that universal treatment was virtually the only way to eliminate the pandemic cost-effectively. While the finely constructed model provided a rationale for further expansion of ART in Africa, it also paved the way for defunding of non-pharmaceutical HIV care, including home-based care activities. The focus on the economic value of intervention and the focus away from the individual state of life is a prominent characteristic of contemporary HIV intervention.

Keywords Cost-effectiveness · Mathematical model · Treatment-as-prevention strategy · Universal treatment

© The Author(s), under exclusive license to Springer Nature
Singapore Pte Ltd. 2023
M. Nishi, *Curing Lives*,
https://doi.org/10.1007/978-981-99-1831-7_3

In 1997, I stayed in a small rural center for several weeks to conduct part of my master course research project (which was not linked to HIV). It was located on an unpaved highway connecting Addis Ababa and the southern part of Ethiopia and was inhabited by several hundred at the time. The town might have looked to have little to offer its visitors, but those familiar with the town could spot places where such items as *aibe* (local cheese), *kibe* (local butter), and *arake* (local liquor) were sold.

Along with the dusty roads, the town had several teahouses and kiosks, the latter displaying a few bunches of *chat* twigs bearing young leaves. *Chat* or *Catha edulis* is a plant grown commonly in the horn of Africa, and its young leaves contain alkaloids that work as a mild stimulant. Visiting one of the town's households in the afternoon, I would come across old and young men gathering to chew chat leaves and enjoy its euphoric effect. When I asked one of those men what it was like to chew *chat*, he explained that it gave him a vivid vision of becoming a millionaire—or he would even feel as if he was already one of them. However, the following morning, he would get up to find himself "being a poor guy." Staying there, I could decline to chew the chat with the men. However, the women in their homes were embarrassed if I were to leave before they offered me a few cups of coffee, the preparation of which began each time by roasting a small amount of coffee beans on burning charcoal.

On one edge of the town stood a small building with a rusting tin roof and walls with half-peeled paint. The signboard on the fence indicated that it was a health post set up by the local government. However, people in town who visited the post often found its doors locked because the only one person stationed there was away for unspecified reasons. When the post was open, it had little to offer its visitors because its small medicine cabinet was almost empty. This was the only government-run health facility supposed to serve tens of thousands of people living in its catchment area.

One who traveled around rural Ethiopia during those days would have come across little, if any, sign of public health investment. This was the result of weak health policies during the past regimes. It was in 1948 that Emperor Haile Selassie's government introduced the Ministry of

Public Health. Until then, a department under the Ministry of Interior had been administering scarce health facilities. Between 1947 and 1972, the number of hospitals rose from 38 to 85, and that of hospital beds from 3300 to 8415 (Stommes & Sisaye, 1980). However, facilities and personnel were concentrated in Addis Ababa and other larger cities. The rural population barely benefited from the emerging health infrastructure.[1]

The Imperial government was replaced by the Provisional Military Government of Socialist Ethiopia (or the *Derg*) after the 1974 revolution. In a WHO report published in 1990, Getachew Tadesse, then the health minister of Ethiopia, wrote that his government supported the concept of primary health care formulated at the Alma-Ata Conference in 1978. He also claimed that the health coverage in the country, which was 15% in 1974, had expanded to 45% (Getachew, 1990). However, Helmut Kloos, a public health expert and close observer of Ethiopia's health policy, noted how the primary health care system failed to function in 1980s Ethiopia: health facilities were underused, health workers lacked community support, and essential drugs were in short supply. The health budget was inadequate,[2] and ambitious health targets were never met (Kloos, 1998). In an article published in a 1988 issue of the *New England Journal of Medicine*, co-authors Richard Hodes, a spine surgeon based in Addis Ababa, and Kloos concluded that "although health services in socialist Ethiopia have undergone major changes since the revolution, they have yet to be shaped into an efficient system, and the health of the population had changed little thus far" (Hodes & Kloos, 1988, p. 923). However, some readers of the article believed the situation deserved stronger words: it was "a disturbing report on the lack of progress in Ethiopia," as the authors of a corresponding letter put it. They went on to claim: "This problem, although certainly exacerbated by internal political strife and natural forces, is not unique to Ethiopia. The lack of access to adequate health care is a global plague and probably represents the most serious barrier to medical progress" (Gannett et al., 1989, p. 1150).

Such legacies of past regimes defined the health system in 1990s Ethiopia. In 1996, Ethiopia had only 87 hospitals (set up by the government or missionary institutions), 243 health centers, and 76 health posts to cover 58 million people scattered across its vast terrain (Federal Ministry of Health, 2005). Moreover, the extent to which they worked as a health system could be contested. In rural Ethiopia, a fatally ill person might be sent to a hospital, provided the family members could afford it. However, those families often did so not in expectation of a cure but to demonstrate their respect and care for the one who was about to die. Hospitals in Ethiopia might have offered symbolic significance for such rural families, and they hardly functioned as part of a system that ensured a longer and healthier life for the nation.

In rural Ethiopia during the 1990s, the therapeutic anarchy that brought much concern among international public health experts did not materialize. Instead, what was found in most parts of rural Ethiopia was the sheer absence of a working health infrastructure. Admittedly, people relied little on it. Once I advised one of my close Ethiopian friends to undergo blood testing when he started complaining of a poor physical condition. Being aware of the amount of sugar he was regularly consuming with coffee and chat (it is common in Ethiopia to use sugar to counter the bitter taste of chat leaves), I was worried that he could develop diabetes. However, he refused my advice, saying, "What if I turn out to be sick?" This was a serious question because he believed the medication provided in local health institutions was poor in quality. Unfortunately, such mistrust could quickly be reinforced by experience. One of his sisters once had persistent hiccups. She visited private clinics in Addis Ababa that were supposed to offer better service than government institutions but did not get well. Several months later, she turned to a traditional healer, who foisted on her an expensive silver ring as the cure.

Who among those familiar with such realities could have imagined in those days that Ethiopia was to have an ART program that would cover much of its vast rural area, work as a system, and save the lives of tens of thousands of people each year?

Transformation of a Health System

Despite all difficulties that worried public health experts, ART worked in many African countries in different settings. Medical anthropologist Susan Whyte (2014) noted that Uganda's ART program was "extremely diverse" (p. 7). At one end, there was the Home-based AIDS Care (HBAC) program, designed to be "purposefully personal, assigning clients a supporter/monitor who followed them into their homes" (p. 8). HBAC was a research site for the US Centers for Disease Control (CDC), with carefully planned interventions. Their clients were all members of The AIDS Support Organization (TASO), an organization for people with HIV established in 1987 and known for its unique support activities. Meanwhile, there were government healthcare facilities, which were often understaffed and ill-equipped in terms of drug supply. In Uganda, institutional healthcare turned out to be "a mosaic of (free) government and (fee) private not-for-profit facilities." However, this did not mean that Uganda's government healthcare system was failing. Instead, as Whyte noted, "there *is* a functioning and heavily used government health care system" in Uganda (p. 9).

Ethiopia's ART system turned out to be different from that of Uganda: it was a coordinated nationwide system controlled by the central government. Those diagnosed as HIV positive in a public or private health institution in Ethiopia would be issued a registration card, locally called *yeketero wereket* (literally, "appointment card"). With this card, they may take their antiretroviral drugs from the institution, always free of charge. Each time they take the drugs, the pharmacist writes on the card the date when they are expected to come back for the replacement. Their ART regimen is codified and specified on the card. The same regimen code is shared throughout the country: in case they run out while away from home, they may take the drugs from a nearby facility, free of charge, simply by showing the card.

The difference between the two countries is partially explained by the difference in the period when HIV interventions started. In Uganda, they started as early as the latter half of the 1980s, partly owing to the community-level movements of patients, their families, and friends (Kaleeba & Ray, 2002). A number of international agencies soon joined the campaign, but they bothered little with "donor coordination." The resulting mosaic of donor-run community-based projects is a familiar landscape (with different contexts) in many African countries.

Nation-wide HIV intervention in Ethiopia started as late as 2005. Small-scale interventions supported by international donors were already in place by the mid-1990s, but society had remained silent. The government and donors were more concerned about recurrent droughts and resulting hunger in rural areas. When Prime Minister Meles Zenawi (who occupied the post from 1995 to 2012) and his government finally recognized HIV/AIDS as a national issue, they soon found low-cost antiretroviral drugs ready for procurement, and donors willing to sit at the table for coordination. Donor coordination has been a common practice in Ethiopia and some other African countries since the latter half of the 1990s.

An official ART program was introduced to Ethiopia in July 2003. Initially, each patient was charged 300–700 ETB (or 36–85 USD at the time) per month, which turned out to be a significant barrier to promoting drug dissemination. The number of patients who started ART by June 2004 was only 9400, whereas the country had around 1 million individuals with HIV at the time. The Ministry of Health targeted 41,000–50,000 new patients on ART for the following years (Federal HIV/AIDS Prevention & Control Office, 2006), but the number was not entirely achievable. Urged by the WHO and UNAIDS, jointly launching the initiative to expand ART for 3 million people worldwide by the end of 2005, the Ethiopian government decided to provide ART free of charge for its poorer citizens.

The free ART program in Ethiopia was planned to operate in "an integrated manner and thereby avoid wastage and duplication of effort" (Federal HIV/AIDS Prevention & Control Office, 2006, p. 17). A coordinating task force was formed in 2004; it was chaired by the Federal Ministry of Health and represented by the WHO, UNAIDS, CDC, and the US Agency for International Development (USAID). Two regional health bureaus (Addis Ababa and Oromiya) and the network association of people living with HIV were also invited to take part in the task force.

The free ART program started in January 2005. However, this was met by another barrier. It required patients to visit local *kebele* office (the lowest level of government administration in Ethiopia) to obtain a "poverty certificate" for free ART (Kloos et al., 2007). Being unable to pay several hundred ETB should not have been a source of shame. However, many patients found it unbearable to have their status exposed to the *kebele* officials, who would not keep the "secret" for themselves.

The same year, the government decided to provide free ART for all its citizens, regardless of economic status.[3] In 2021, 610,000 people were estimated to be living with HIV in Ethiopia, of whom 480,000 were on ART. Ethiopia became one of the biggest recipients of the Global Fund, which invested 2.75 billion USD from its establishment in 2002 until 2022.[4] HIV incidence per 1000 population dropped from 3.13 in 1992 to 0.12 in 2021. It is estimated that 270,000 deaths were averted due to ART in Ethiopia between 2005 and 2016.[5]

The rollout of ART in Ethiopia coincided with a period of aggressive investment in the public health sector when Dr. Tedros Adhanom (who later became the Director-General of the WHO) served as the health minister between 2005 and 2012. In 2013, Dina Balabanova and colleagues described Ethiopia, along with Bangladesh, Kyrgyzstan, Thailand, and the Indian State of Tamil Nadu, as having "either achieved substantial improvements in health or access to services or implemented innovative health policies relative to their neighbours" (Balabanova et al., 2013, p. 2118). Indeed, Ethiopia's health success is not limited to HIV intervention. It took as long as 34 years until 1997 for Ethiopia's national life expectancy to gain ten years from 40 to 50 years. However, the figure improved to 60 years in only 12 years between 1997 and 2009, and reached 67 in 2020.[6] The country's infant mortality rate halved from 117 per 1000 live births in 1991 to 54.3 in 2010, closing the gap with India, whose number was 45.1 in 2010.[7] Ethiopia's health gain was facilitated by the rapid expansion of its rural health infrastructure. The number of health centers in Ethiopia increased from 243 in 1996 to 3547 in 2015 (Federal Ministry of Health, 2015, p. 49). When I visited in 2009, a health center located 50 km off the main highway had several units dedicated to such health issues as mother–child health, TB, and HIV. It was run by a dozen health professionals with various qualifications. Its pharmacy was stocked with essential medicines.

Balabanova and colleagues (2013) attributed Ethiopia's success primarily to the political leadership within the country. They named Meles Zenawi, the then prime minister of Ethiopia, and Tedros Adhanom as the prominent sources of the leadership. Led by these distinctive politicians, the government pursued "a coordinated program including many development partners, with a focus on access to essential services in isolated areas, based on an innovative health extension program" (p. 2121). In 2003, the Ethiopian government introduced the health extension program under which it deployed more than 30,000 health extension workers (community health workers) who delivered key health messages to rural communities (Wang et al., 2016). The health gain registered among the population during the following years was one of the essential factors that led to Dr. Tedros being elected as director-general of WHO in 2017.

A Mathematical Model

Ethiopia's success out of a near-vacuum of health infrastructure proved that a working ART system could be installed in an African country. The process was much faster and cheaper than some public health experts initially anticipated. Most states in Africa developed a less coordinated system than Ethiopia, but the ART coverage in Africa turned out to be generally successful. It occurred in the absence of the health infrastructure initially demanded by experts.

Abundant data from Africa helped epidemiologists make sense of what they were seeing. One of the early responses by pro-treatment experts was to confirm that "therapeutic anarchy" did not prevail in Africa. In their study, epidemiologist Edward Mills and colleagues (2006) compared adherence to ART in North America and Africa by performing a meta-analysis of 58 studies (of which 31 studies represented North American populations and 27 represented the populations of 12 African countries). An interesting result was that Africans were doing better: an estimated 55% of the American populations achieved adequate levels of adherence, whereas the African studies indicated a higher adherence, at 77%.

The study pointed to both settings having patients with sub-optimal adherence, and indicated that concerns about sub-optimal adherence "should not contribute to delayed access to treatment" (Mills et al., 2006, p. 685). It could also be said that there was no such thing, in the first place, as a system to ensure optimal adherence, even in North America. In 2012, Abigail Zuger, a medical doctor at Mount Sinai Roosevelt and Mount Sinai St. Luke's Hospitals in New York City, observed the reality as follows:

> In our busy urban clinic, some patients are as precise with their medications over years and decades as anyone might wish. The rest offer up a true symphony of reasons for nonadherence. Some are too depressed, distracted, drug-addled, disengaged, blasé, or suspicious to take their medications consistently. Some sell the drugs on the black market, and some share them with friends and partners. Some work out their own schedule of treatment interruptions, contrary to all medical advice. One patient of mine, a nurse, actually took half a regimen for a solid calendar year. The upshot: The pharmacokinetics of that regimen made him a very lucky guy, and he had an undetectable viral load at the end of that year. Another patient was less fortunate: She couldn't get to her refills for a month—and her virus could never be controlled again. (Zuger, 2012)

Zuger's observations provide a vivid explanation of how medication adherence could be anarchic, to a varied extent, in different parts of the world.[8]

Using the rich data extracted from the ART experience in Africa, epidemiologists came up with more evidence to encourage universal treatment and ideas for more effective public health interventions. Their knowledge was soon crystallized as the "treatment as prevention" (TasP) strategy, which assumed ART was a powerful tool for eliminating viral transmission. Successful ART not only prolongs the patient's life span but also significantly reduces the likelihood of the virus being transmitted to others because it substantially lowers the patient's viral load, often to an undetectable level. While this was a clinically proven fact, its epidemiological significance was yet to be fully understood: epidemiologists were eager to verify whether universal ART in Africa and elsewhere in the world would lead to the swift elimination of the HIV pandemic. Two groundbreaking studies suggested it would do so.

Based on data from South Africa, Reuben Granich and colleagues (2009) constructed a mathematical model to simulate the long-term dynamics of the HIV epidemic under different intervention scenarios. The study demonstrated that, in a hypothetical community, the epidemic could be brought to near elimination within ten years under a universal treatment scenario. In this scenario, they assumed universal voluntary HIV testing (all individuals older than 15 years are offered voluntary HIV testing once a year) and immediate ART (everyone diagnosed as HIV positive is put under treatment immediately). To make their model more realistic, they assumed that a particular portion of people would refuse treatment or fail to adhere to it, based on data from the national ART program in Malawi. This hypothetical model was supported by the outcome of a large-scale clinical trial named the HIV Prevention Trials Network 052 study. This was a Phase III, two-arm, randomized, controlled, multicenter trial started in 2005 to determine whether ART could prevent the sexual transmission of HIV in serodiscordant couples (in which one partner is HIV positive and the other negative). The study was conducted in nine countries, of which five were African, and enrolled 1763 couples. They were randomly assigned at 1:1 ratio to receive ART either immediately or after a decline in their CD4 count or the onset of HIV-related symptoms. (Note that such delayed treatment intervention was considered standard practice at the time. The advantages of delaying treatment include avoiding adverse effects.) The results pointed to the clear advantage of immediate treatment in preventing new transmissions: only one HIV transmission was reported in the immediate treatment group, whereas 27 transmissions were observed in the delayed treatment group (Cohen et al., 2011).

Besides its powerful effect on eliminating new HIV transmissions, another suggested advantage of the TasP strategy is its cost-effectiveness. Implementing immediate ART increases the number of people on treatment worldwide, leading to a surge in global investment in HIV in the coming years. Granich and colleagues calculated, based on their mathematical model, that the cost of treatment would soon start to decrease,

as the number of new infections drops due to universal and immediate ART. Thus, the long-term cost of global ART would be cheaper
than the conventional intervention (Granich et al., 2009). With such
evidence, epidemiologists are more confident than ever that the TasP
strategy should be at the center of efforts to eliminate the HIV pandemic.

Defunding of Non-Pharmaceutical Care

One of the consequences of focusing on the cost of treatment was
defunding of non-pharmaceutical care activities in Africa and elsewhere in the Global South. In Ethiopia, such activities were primarily
conducted by self-help associations of people with HIV. Some began
operations during the 1990s, representing crucial aspects of HIV care
work and relationships in the country while providing social and material support, including peer counseling, home-based care, and food and
monetary assistance for their members.

The proliferation of HIV self-help activities coincided with the early
stage of ART scale-up in Ethiopia, as I will elaborate in Chapter 4.
Between 2004 and 2009, many such associations were established
countrywide and encouraged via moral and financial support from
transnational NGOs and funding agencies. The national universal ART
program, coupled with non-pharmaceutical programs conducted by self-
help associations, once seemed to comprise an ideal state–civil society
partnership in HIV care.

The situation changed significantly following the global financial
crisis triggered by the bankruptcy of Lehman Brothers in 2008. Serieux
and colleagues (2012) reported that in Malawi, the national economy
survived the crisis relatively unscathed, but there was a sharp deteri-
oration in funding for agencies engaged in HIV programs. The same
happened in Ethiopia. While the country's economy continued to grow
at relatively high rates during the consecutive years, international HIV
funding shrunk. The situation was worsened by a shift in state policy
toward civic organizations. In 2009, the government of Ethiopia issued
a proclamation that severely restricted the activities and resource access

of non-governmental organizations, including the associations of HIV-positive people (see Chapter 4).

Since then, it has become increasingly evident that non-pharmaceutical care activities are underfunded and activities of HIV self-help associations marginalized. Between July 2011 and June 2012, 125 million USD were spent on HIV treatment and care programs, of which 69 million (55%) were spent on ART, according to data compiled by the Ethiopian government (Federal HIV/AIDS Prevention & Control Office, 2013). In contrast, only 1.4 million (1.1%) went to home-based care, a pillar of self-help association activities.[9] Kalofonos reported that, in Mozambique, home-based care activity was incorporated into the national program and was redefined to fit a scheme that narrowly defined care (2021). In Ethiopia, it was not redefined but was defunded and deserted.

Continuity in the Global Health Governance

The universal HIV treatment (or the TasP) strategy is an unprecedented health project that calls for a novel assemblage of knowledge, technologies, and institutions. Its rollout in sub-Saharan Africa accompanied significant shifts of focus in public health interventions in at least two ways: from the moral approach to universal treatment; health infrastructures to mathematical models. However, there is substantial continuity as well. The TasP strategy, I argue, is a recent addition to the continuum of biopolitical experimentations aiming at governing the population in the Global South. The focus on the economic value of intervention and the focus away from the individual state of life is an inherent characteristic of the tradition. Two historical accounts of the comparable practices are helpful to elaborate on my point: Michelle Murphy's study of the "economization of life" promoted by family planning in Bangladesh (2017) and Keith Breckenridge's discussion of the emergence of a distributive state in South Africa in the absence of an elaborate bureaucracy (2014).

Murphy evokes the history of family planning in the 1970s as crucial to the emergence of global health experimentality in Bangladesh and elsewhere in the developing world: "HIV care in the 1990s was often initially assembled through already existing infrastructures of "reproductive health" care, in turn built out of the family planning infrastructures of the 1970s" (2017, p. 80). This observation seems to apply to the case of Ethiopia, as indicated in the recollection of a traditional birth attendant in Addis Ababa (Chapter 7).

However, how some tenets of twentieth-century population control programs were passed down to the present-day HIV interventions is not straightforward. Murphy argues that family planning has attached negative economic value to the possible life of future people, particularly poor people in the Global South, by enforcing its central assumption that "some must not be born so that future others might prosper" (p. 46). This assumption is reminiscent of the similar tenet that inhibited the rollout of ART in Africa: those in poor countries must not be treated so that the richer world would be protected from the anticipated "antiretroviral anarchy" in Africa. As I discussed above, the TasP strategy represents a departure from the necropolitical assumption underpinning past interventions because it assumes that everyone, regardless of race and economic status, must be treated to end the HIV pandemic. Nevertheless, as I will argue in the following chapters, the rollout of the free ART program resulted in what Murphy referred to as the "economization of life," that is, the systematic practice of designating lives as worthy of investment and lives that are not.

Breckenridge's study on the colonial origins of a distributive state in South Africa illuminates how a form of biopolitical governance persists despite dramatic political transformations. He elaborated on how the method of biometric surveillance—using fingerprints—developed during the colonial and the Apartheid eras turned out to be the vital technology for expanding cash transfer programs, such as the Child Support Grant, by the post-Apartheid state (2014).

The effort to capture and classify the fingerprints of African subjects started in the early twentieth century. The health of the native population, who were prone to the epidemics of malaria and tuberculosis, was a primary concern of the proponents of the biometric measure. The Transkei division of the South African Medical Association (SAMA), based in the middle of the largest native reserve, was among the keen promoters of native registration. In one of their letters addressed to the Secretary of Native Affairs,

> the doctors drew out the embarrassing comparison between the state's expectations for livestock and the requirements for people: 'If it is found necessary to register the Births and Deaths of cattle, it would appear that the registration of the Births and Deaths of the people should be even more necessary.' They cited the data from a single hospital in the Transkei to show the horrible effects of tuberculosis in the countryside—over 20 percent of total deaths were derived from this single cause. 'Reliable statistics,' they argued in a statement that echoes through to the recent HIV crisis, 'would make the need for a [national health] service abundantly clear.' (Breckenridge, 2014, p. 132)

Indeed, the focus on population and statistics and the focus away from personal state of life is reminiscent of the recent HIV intervention, despite significant differences in the historical context and transformation of the ethical premise.

Moreover, the economic value of the intervention dominated the policy negotiation. Despite the health experts' insistence on civic registration, the colonial government was reluctant to spend money on the native population. The Secretary of the Interior rejected the idea stating that the previous efforts to register native births and deaths proved very unsatisfactory. Furthermore, he argued that the resulting registration of Africans was of "small economic value" and the wisest policy was to suspend registration in the countryside until some future when "the native may be sufficiently advanced in the scale of civilization to realise the advantages of registration." He concluded that it would not be acceptable to the Department of Interior unless "it could be conclusively shown that such a scheme would be of economic value and of real interest from a health point of view" (Breckenridge, 2014, p. 130).

The effort for civic registration in South Africa was later revitalized under the National Party's government, which saw fingerprint registration as a useful tool to implement its segregation policies. In a twist of history, this scheme turned out to be a vital infrastructure of "biometric citizenship" for the post-Apartheid government to develop a universal system of social benefits. The centralized biometric register "made possible the rapid and very wide delivery of the cash payments for child support and pensions that are being acclaimed around the world," for this is an "extremely efficient and highly scalable technology which underpins the current interest in a global system of cash grants for the poor" (Breckenridge, 2014, p. 137).

What outlived the significant political transition in South Africa was the obsession with efficiency and the absence of interest in the individual citizen's state of life. This is not to say that South Africa's present distributive schemes are morally dubious because they use biometric technology that has its root in a colonial polity.[10] My point here is that some tenets of present-day HIV intervention, particularly the focus on economic value and the focus away from the personal state of life, are not recent inventions. On the contrary, they are long-standing doctrines of health governance in the Global South.

Change and Continuity

The rollout of universal HIV treatment in Africa has been an unprecedented experiment that turned the continent into a "living laboratory" (Tilley, 2011), simultaneously testing social justice and the health of a population. The epidemiological knowledge that underpinned such an endeavor was extracted from the ART experience in Africa (Granich et al., 2009). It facilitated a significant shift in the moral basis for global HIV intervention, at least in two ways.

First, it swept away the argument for the so-called moral approach that effectively denied treatment access to people with HIV in low-income countries by insisting that "behavior change" was the only way to avoid contracting the virus and consequent deaths. The moral approach essentially operated on the theoretical premise of individual self-discipline: its proponents insisted that HIV transmission could be eliminated if all members of a community observed abstinence and faithfulness, which never happens in the real world. It also operated upon a paternalistic moral framework—"We know what is good for Africans"—that covered up the fundamental bigotry that it was "moral" to allow people with

HIV in low-income countries to die while their counterparts in the high-income world should live. With the epidemiological knowledge that powerfully supported universal treatment, the global HIV intervention departed from the moral approach, which had been dominant during the 1990s.

Second, and more importantly for my argument, the rationale for universal ART departed from humanitarian moral grounds as well. In the years preceding the rollout of ART in Africa, Zackie Achmat, a former anti-Apartheid activist, refused to take antiretroviral drugs until all who needed them had access. Members of the Treatment Action Campaign, the most prominent HIV advocacy group in Africa, and their supporters were marching the streets with banners and placards that demanded universal treatment.[11] After the global rollout of ART, the placards were replaced by mathematical models, and popular demonstrations by clinical trials. Universal treatment is now embraced by health experts and policymakers as a public health strategy for safeguarding the health of the populations by eliminating a pandemic most swiftly and cost-effectively.

The focus on the economic value of intervention and the focus away from the individual state of life is a long-standing tradition of health governance in the Global South (Breckenridge, 2014). These are the tenets that outlived the shifts in the moral basis of the global HIV intervention. The TasP strategy, in this regard, is a recent addition to the continuum of biopolitical experimentations aiming at governing the African population.

It is not my intention, however, to indicate that the TasP strategy itself was deemed or intended to alienate humanity or promote the economization of life. As Murphy (2017) put it, there are many ways of doing an experiment, and more crucially, an experiment has the potential "for generating life otherwise, for yearning toward the possibility of other worlds and other arrangements that might be less violent and more affirming to life" (p. 81). In the course of the global fight against HIV during the past decades, the human ability to imagine a world otherwise made way for universal treatment. Nevertheless, the prevailing HIV intervention in the Global South falls short of "affirming to life." It requires a closer investigation to figure out how it happened.

The rest of my book offers a closer look at some of the effects of such an evidence-based, cost-effective approach to ensuring longer and healthier lives for the African population. It traces, through the eyes of an Ethiopian woman whose name is Meseret, the course of HIV movements in Ethiopia since 1997. By doing so, it reveals an ambivalent landscape of therapeutic citizenship in the times of universal treatment.

Notes

1. In 1972, 92% of the expenditure on medical care covering all sources was on hospitals. The two largest cities, Addis Ababa and Asmara, spent 62.4 ETB and 34.5 ETB per capita, respectively, compared to 2.9 ETB for the rest of the country (Kloos, 1998, p. 507).
2. The proportion of the national budget spent on health services declined from 6.1% in 1973/1974 to 3.5% in 1986/1987 and further to 3.1% in 1990/1991. The annual increase in the capital health budget was slightly over 1% instead of the planned 11% (Kloos, 1998).
3. Initially, patients with compromised immune systems (CD4 count less than 200 cells/mm^3) were eligible for the national ART program. However, the clinical criterion was revised in 2012 to include patients with a CD4 count less than 350 cells/mm^3 and in 2014 to include all adults with a CD4 count less than 500 cells/mm^3, adults with active TB disease, hepatitis co-infection, pregnant and lactating women, and children less than 15 years of age (Assefa et al., 2017). Since 2016, all individuals diagnosed with HIV have been eligible for the free ART service regardless of their CD4 counts. As per the national guidelines, rapid ART initiation is offered to all individuals within seven days of a confirmed HIV diagnosis (Federal Ministry of Health, 2018).
4. Ethiopia is the third-largest recipient of the Global Fund after Nigeria (3.25 billion USD) and Tanzania (2.86 USD). The figures represent the total disbursements by the Global Fund to projects covering HIV, malaria, and tuberculosis. Data obtained from the Global Fund Data Explorer, https://data.theglobalfund.org/viz/disbursements/map. Accessed on October 28, 2022.
5. Data extracted from Our World in Data, https://ourworldindata.org/grapher/hivaids-deaths-and-averted-due-to-art. Accessed on October 28, 2022.

6. Data extracted from World Bank Open Data, https://data.worldb ank.org. Accessed on October 28, 2022.

7. Data extracted from World Bank Open Data, https://data.worldb ank.org. Accessed on October 28, 2022. The health gap between India and Ethiopia is peculiarly narrow, considering the huge gap in the infrastructure and human resources between the two countries. One of the factors for India's poor health outcome is the huge inequities in access to health care (Bhalotra, 2007; Duggal, 2007).

8. The higher-than-expected adherence in Africa remains largely unexplained, as there are too many factors that may affect a patient's medication behavior. However, Mills and colleagues (2006) indicated that the complex ART regimen in North America at that time might have affected the adherence of its population. To simplify the ART regimen, African systems widely used fixed-dose combination drugs. A combination drug usually contains in a tablet three different types of antiretroviral drugs to achieve viral suppression. While fixed-dose combination drugs have the obvious disadvantage of being unable to accommodate personalization of the ART regimen for each patient to achieve higher viral suppression and/or lower side effects, they offer a lower adherence burden to the patient. Such drugs were first introduced by Cipla, the multinational pharmaceutical manufacturer based in India, targeting low- and middle-income markets.

9. Other treatment and care spending included outpatient care services (12%), laboratory monitoring (11%), and provider-initiated testing and counseling (9%). HIV expenditures outside the treatment and care category included 79 million USD on prevention activities, 120 million on national system strengthening and program coordination, and 10 million on social protection services such as income generation activities (Federal HIV/AIDS Prevention and Control Office, 2013). Limited data were available concerning the spending trends during the following years. Another data set shows that total HIV spending in Ethiopia decreased by half between 2011 and 2019. In 2019, 52 million USD or 44% of HIV treatment, care, and support spending was on ART. Home-based care expenditure is not indicated. Data obtained from UNAIDS HIV Financial Dashboard, https://hivfinancial.unaids.org/hivfinancial dashboards.html. Accessed November 16, 2021. Note that spending categories are inconsistent among the data sets.

10. Ferguson argued against an overly pessimistic perspective on emerging distributive experimentations in South Africa and elsewhere in the

Global South (2015). He viewed cash grants as affirming the idea of "rightful share" or "a binding entitlement" of the poor "to some specified share of the total global production" (p. 38).

11. *Fire in the Blood*, a 2013 film directed by Dylan Mohan Gray and which starred Zackie Achmat, presents vivid images of the HIV movements in South Africa.

References

Assefa, Y., et al. (2017). Performance of the antiretroviral treatment program in Ethiopia, 2005–2015: Strengths and weaknesses toward ending AIDS. *International Journal of Infectious Diseases, 60*, 70–76.

Balabanova, D., et al. (2013). Good health at low cost 25 years on: Lessons for the future of health systems strengthening. *The Lancet, 381*(9883), 2118–2133.

Bhalotra, S. (2007). Spending to save? State health expenditure and infant mortality in India. *Health Economics, 16*(9), 911–928.

Breckenridge, K. (2014). *Biometric state: The global politics of identification and surveillance in South Africa, 1850 to the present*. Cambridge University Press.

Cohen, M. S., et al. (2011). Prevention of HIV-1 infection with early antiretroviral therapy. *The New England Journal of Medicine, 365*(6), 493–505.

Duggal, R. (2007). Healthcare in India: Changing the financing strategy. *Social Policy and Administration, 41*(4), 386–394.

Federal HIV/AIDS Prevention and Control Office. (2006). *Report on progress towards implementation of the declaration of commitment on HIV/AIDS*. Federal HIV/AIDS Prevention and Control Office, Federal Democratic Republic of Ethiopia.

Federal HIV/AIDS Prevention and Control Office. (2013). *Ethiopian national AIDS spending assessment report EFY 2004, 2011/12*. Federal HIV/AIDS Prevention and Control Office, Federal Democratic Republic of Ethiopia.

Federal Ministry of Health. (2005). *Health sector development plan III*. Federal Ministry of Health, Federal Democratic Republic of Ethiopia.

Federal Ministry of Health. (2015). *Health sector transformation plan*. Federal Ministry of Health, Federal Democratic Republic of Ethiopia.

Federal Ministry of Health. (2018). *National consolidated guidelines for comprehensive HIV prevention, care and treatment*. Federal Ministry of Health, Federal Democratic Republic of Ethiopia.

Ferguson, J. (2015). *Give a man a fish: Reflections on the new politics of distribution*. Duke University Press.

Gannett, P., et al. (1989). Correspondence. *New England Journal of Medicine, 320*(17), 1150.

Getachew, T. (1990). Ethiopia: The course is charted. In E. Tarimo & A. L. Creese (Eds.), *Achieving health for all by the year 2000: Midway reports of country experiences* (pp. 80–97). World Health Organization.

Granich, R. M., et al. (2009). Universal voluntary HIV testing with immediate antiretroviral therapy as a strategy for elimination of HIV transmission: A mathematical model. *The Lancet, 373*(9657), 48–57.

Hodes, R. M., & Kloos, H. (1988). Health and medical care in Ethiopia. *New England Journal of Medicine, 319*(14), 918–924.

Kaleeba, N., & Ray, S. (2002). *We miss you all*. SAfAIDS.

Kalofonos, I. (2021). *All I eat is medicine: Going hungry in Mozambique's AIDS economy*. University of California Press.

Kloos, H. (1998). Primary health care in Ethiopia under three political systems: Community participation in a war-torn society. *Social Science & Medicine, 46*(4), 505–522.

Kloos, H., et al. (2007). Utilization of antiretroviral treatment in Ethiopia between February and December 2006: Spatial, temporal, and demographic patterns. *International Journal of Health Geographics, 6*, 45.

Mills, E. J., et al. (2006). Adherence to antiretroviral therapy in sub-Saharan Africa and North America: A meta-analysis. *Journal of the American Medical Association, 296*(6), 679–690.

Murphy, M. (2017). *The economization of life*. Duke University Press.

Serieux, J. E., et al. (2012). The Impact of the global economic crisis on HIV and AIDS programs in a high prevalence country: The case of Malawi. *World Development, 40*(3), 501–515.

Stommes, E., & Sisaye, S. (1980). The development and distribution of health care services in Ethiopia: A preliminary review. *Canadian Journal of African Studies, 13*(3), 487–495.

Tilley, H. (2011). *Africa as a living laboratory: Empire, development, and the problem of scientific knowledge, 1870–1950*. University of Chicago Press.

Wang, H., et al. (2016). *Ethiopia health extension program: An institutionalized community approach for universal health coverage*. The World Bank.

Whyte, S. R. (Ed.). (2014). *Second chances: Surviving AIDS in Uganda.* Duke University Press.

Zuger, A. (2012). A skeptic looks at 'test and treat.' *Journal Watch HIV/AIDS Clinical Care.* https://search.proquest.com/scholarly-journals/skeptic-looks-at-test-treat/docview/1319247515/se-2?accountid=11929.   Accessed   on March 13, 2021.

4

In Search of a Cure

Abstract This chapter offers an account of the state of life before ART through the eyes of an Ethiopian woman whose name is Meseret. After she lost her husband to AIDS and was diagnosed as HIV positive herself, Meseret was subjected to constant harassment by her in-laws, so much so that she finally decided to leave their village. She sought refuge in Entoto Mountain, where thousands of people "like her" lived in shacks on the mountainside. However, the time of silence was followed by a moment of extraordinary change when people with HIV and their associations rapidly acquired a public presence. Meseret left Entoto Mountain and returned to her home province to establish a small association of people with HIV. These were formative years for Ethiopia's HIV activists when they departed from where they once were. I suggest that this departure was significant in two ways: it meant a departure from a life lived in fear and silence, and it also meant a departure from close engagements with the suffering of others. However, keeping some distance from the national movement allowed Meseret to continue to engage closely with the problems of others like her.

Keywords Advocacy · Fear · HIV movement in Ethiopia · Silence

I first met Meseret in September 2007 in Welkite, the capital of the Gurage Zone, which is located 150 km southwest of Addis Ababa. The Gurage Zone is one of the administrative units in the Southern Region of Ethiopia.[1] The name derives from the ethnic Gurage people who mainly occupy the area. Its rural areas are inhabited primarily by ethnic Gurage peasants, whereas Welkite is a commercial town with mixed ethnic populations. I visited the town hoping to start my new research project concerning local responses to the HIV epidemic. During a visit to the office of an NGO based in Welkite, I saw a poster that promoted HIV screening. The poster included a photograph of a woman with HIV as well as her message: "HIV screening made it possible for me to know myself and to consider what to do for my life."

The woman on the poster was Meseret. Our first interview took place in her office. I told her I saw the poster and was impressed by the positive message. She said, "Let me tell you what I encountered. Some people who saw my poster hanging on a wall tsk-tsked [to show pity], saying, 'This woman could have been dead.'" Free ART was already available in Ethiopia at the time, but many continued to perceive AIDS as a fatal illness. However, Meseret did not seem embarrassed by the story. Her tone indicated that she was not blaming the ignorance of those people. It was as if she talked about a funny encounter that could happen in anyone's life.

Meseret was in her mid-20s and served as the representative of a small association of people with HIV living in and around the town. On my third visit, in August 2009, she shared some more pieces of her "funny" experience. Once, she traveled with one of her colleagues to Butajira, another town in the Gurage Zone, where they attended an HIV conference. When they were having dinner at the hotel they stayed in, a well-dressed middle-aged man approached them and asked if he had met them some place. "Some place is not clear enough. Where did we meet?" asked Meseret. He said he did not remember. When the two finished dinner and tried to retreat to their room, he followed them and insisted that they had to have some tea together. They went back to the restaurant. Meseret ordered a Harar Sofi (a nonalcoholic malt beverage popular in Ethiopia) while her friend had a beer. After a while, the man was visibly drunk and started to insist that he wanted to sleep with one of them. Meseret asked him if he was using condoms. He answered that he was married, implying that he was a "faithful" man and, therefore, did not need to use one. When she finally told him she had HIV, he went pale, said nothing, and hastily retreated to his room. She had more or less expected that response. The following day, Meseret and her colleague roamed the hotel courtyard to give him a chance to explain himself. The man finally came out of his room and apologized that he had offended Meseret and her friend.

The man's self-contradictory remark that he was "faithful" and therefore safe to have unprotected sex was not unique to him but was typical of the widespread male response to the HIV risk overshadowing their construction of sexuality and masculinity. Anthony Simpson (2009) found, in his ethnographic research with a cohort of Zambian men whose adolescence and adulthood coincided with the HIV crisis, that most of them engaged in unprotected extramarital sex. These men educated at a Zambian Catholic mission boarding school were well aware of the risk involved in sexual intercourse and were convinced they loved their wives. However, they were also under constant pressure to prove themselves to be "real men," and they assumed they could do so by engaging in extramarital sex.

In Ethiopia, too, men tended to bear such contradictory perceptions, and as a result, it was a common experience for young women to encounter married men who claimed to be "safe" to have unprotected sex. What made Meseret's story unique was her ability to counter the injustice committed against women and people with HIV. Admittedly, it is not reasonable to expect every woman to act in the same manner because in Ethiopia (like other parts of the world), it is often difficult to convince men of their wrongdoing, and attempting to do so may trigger violence. However, at the same time, it is unreasonable to assume that Meseret obtained such a capacity out of nothing. I will argue later in Chapter 7 that her modes of action can be related to what I refer to as the "culture of defiance" shared among some Ethiopian women.

Before returning to Meseret's story, let me sketch some more aspects of the epidemic in Ethiopia compared to other African societies. The country has seen a relatively low level of HIV infection compared to other parts of Eastern and Southern Africa, where the prevalence has been by far the highest in the world.[2] In Ethiopia, HIV prevalence among adults peaked at 3.0% in 1995, while the figure reached 9.6% in Uganda in 1990 and 14.7% in Zambia in 1998. The number of people living with HIV in Ethiopia was 0.61 million (out of 122 million) in 2021, compared to 1.4 million (out of 47 million) in Uganda and 1.3 million (out of 19 million) in Zambia in the same year.

By referring to these numbers, I am not trying to indicate that Ethiopians had less severe problems. However, a relatively small number of infections in a given society provides more psychological space for people to perceive that HIV is "other people's problem." I once heard one of my Ethiopian friends say, "I wonder why AIDS patients do not die quickly before infecting others." None of his family, close relatives, or best friends seemed to have died of AIDS by that time. I wondered if he would have said the same if he had lost someone close to him.

It is often said that in Africa, HIV has affected everyone regardless of the infection status. This remark would have sounded more relevant in Zambia and Uganda, where virtually every member of the society had lost some of their family members or close friends to AIDS. However, the illness was not perceived the same way in Ethiopia's public spaces. On December 7, 2000, I was among the audience who sat in the Africa Hall and listened to the speeches of African leaders at an international conference dedicated to discussing the challenges of HIV/AIDS.[3] In his remarks, Ethiopian prime minister Meles Zenawi referred to the disease as the "deadly enemy of our continent," which sounded unsympathetic to more than 18 million people who were estimated to be living with HIV in the continent in the same year. His tone contrasted with the words of Yoweri Museveni, the Ugandan president, who remarked on the same occasion, "People living with AIDS need love, care, and understanding like everybody else." Leaders might not always act according to their words. Still, their words are important because of their influence over citizens' attitudes toward those affected by the epidemic.

Another significant issue is the higher prevalence of HIV among women. In 2021, Ethiopian women were 2.3 times more likely to be infected than men, and the gap had widened over time. A similar trend was observed throughout Africa. Despite the significant difference in prevalence levels, African societies demonstrated striking consistency in that women were more vulnerable to HIV infection (Birdthistle et al., 2019). It is difficult to figure out why this happened because how gender interplays with the epidemic is a complex issue: Research points to multiple biological, social, behavioral, cultural, economic, and structural factors in African societies (Ramjee & Daniels, 2013). Furthermore, social and economic factors contribute to the distress of women with HIV. Stigma against the disease and women's weak claim to the land and other assets resulted in the deprivation of women who lost their husbands to AIDS, as we will see in the case of Meseret in the following section of this chapter.

Diagnosis

"Thanks to AIDS, I reached where I am now. Otherwise, I was an ordinary rural woman," Meseret told me in our first interview. It was only later that I learned that such a simple remark could not express the course of life she had gone through. She was born in a village some 70 km away from Welkite. She was married to a village man in her early teens but was divorced after her first baby died at two months old. She married again, this time by her own will, to an ex-military man. Shortly after she gave birth to a baby boy, the war against Eritrea erupted.[4] Her husband reenlisted in the armed forces and left for the war front in the desert area close to Assab, the Eritrean port on the Red Sea. He came home in 1999 and stayed with her for about a month before returning to duty. It was at that time when one of Meseret's friends saw him and said, "Your husband lost weight, and his face became darker. What happened to him?" Meseret answered, "Didn't he stay in the desert? That should be the reason," to which the woman responded, "Good, if so."

After her husband left home, Meseret remembered those words and became very anxious: Was the friend trying to indicate, by those obscure words, that he could have HIV? Then it came to her mind that she could consult her brother, whom she had not seen for years. He was among a batch of teenagers sent to Cuba—a close ally of Socialist Ethiopia until the latter's fall in 1991—and trained there to become a medical doctor. He had recently returned to Ethiopia and worked at the hospital in Jimma, a town 200 km away from Meseret's village. She tried to call him on the phone, but the line was always too noisy. After several attempts, she finally talked to her brother, confirmed his address, and immediately left for Jimma with her son. However, when she met him, she did not know how to explain her situation. After several days, the brother noticed her uneasiness and asked her to explain. She told him that she needed to get tested for HIV. When she got tested at the hospital, the result was negative, but she was told to come back three months later for another test to confirm her HIV status.

Meseret returned home to find her husband had come back two days earlier, seriously sick. He insisted he had contracted malaria and took the antimalarial drugs prescribed at the local health center. When his condition further deteriorated, Meseret persuaded him to go to the Catholic hospital near Welkite. After examining his blood, the doctor told Meseret's husband that he had a "problem which is not malaria," indicating that he tested HIV positive. Similar to other African countries at the time, HIV screening was available at local hospitals and clinics, but treatment was unavailable. On hearing the words, "I stretched my arms in the air and fainted away," recalled Meseret. Her reaction surprised the hospital's medical staff, who did not realize until that moment that she was the patient's wife. She was so young that they had taken her as his sister.

When they returned home, she did not tell her in-laws that her husband had tested HIV positive. One of his sisters suspected that his sickness was caused by "a certain [malicious] deed that had been done on him" and insisted that they had to consult the witch doctor. When Meseret rejected the idea, the sister was annoyed and insulted her by saying, "You are not troubled. In case he is dead, you will marry another." The sister then went to see the witch doctor, who told her the kind of sheep to be sacrificed. When she bought the sheep and asked Meseret to pay for it, Meseret did so very reluctantly. Not long after the sheep was slaughtered, her husband died. Meseret went through additional tests and was confirmed to be HIV positive.

Like many African widows who had lost their husbands to AIDS, her status in the village was vulnerable (Izumi, 2007; Swaminathan et al., 2008). Geissler and Prince (2010) noted that, among the Luo people in Kenya, the death of a man was often followed by competition over the land, which created a "battleground about gender, kinship and ownership, upon which widows hold a precarious position" (p. 273). The same happened to Meseret. In the hopes of driving Meseret away from her deceased husband's land, her in-laws became increasingly hostile toward her.

For the Gurage people, land is a scarce resource that underpins prosperity and family continuity; consequently, the rural Gurage population is plagued by frequent land disputes. Moreover, women's property rights have been a thorny issue in the Gurage Zone and elsewhere in Ethiopia. Access to land is construed by customary and modern laws, which often contradict each other. Gurage customary law has been interpreted and enforced by male elders unwilling to recognize female land rights. In 2000, the federal government proclaimed the Revised Family Code to reinforce female rights in marriage and their access to property (Federal Democratic Republic of Ethiopia, 2000). Still, it fell short of helping widows like Meseret for several reasons—after all, it was unthinkable for an "ordinary rural woman," as she was at that time, to stand against local elders who governed the community.[5] However, the customary law entitled Meseret's son to claim access to his deceased father's land. That was why Meseret's in-laws were determined to harass her so that she would not stay there until the son was old enough to stand in front of the local elders to confirm his access to the land. "I wanted to cling on to the land for the sake of my son," recalled Meseret. "But finally, I decided to leave the village."

Recalling those days, she said, "I was in despair at the time. It seemed to me that someone with HIV would quickly die." Like many other individuals with HIV in Ethiopia, Meseret found herself in despair and isolation. However, in Addis Ababa, connections among people with HIV were already forming.

Prologue

Established in 1997 in Addis Ababa, the Mekdim Ethiopia National Association (hereafter, Mekdim) claims to be the first association of people with HIV in Ethiopia. Its creation resulted from international and local connections that shaped the HIV movements in the country. Its formation was primarily inspired by Ugandan civic activities, which have been active since the mid-1980s in providing care and support for those dying of AIDS. At the same time, Ethiopia's sociocultural milieu, particularly since 1997, shaped the association's particular version of therapeutic citizenship.

In August 2015, I visited Mekdim's headquarters, which occupied a strip of the plot facing an alley adjacent to the Menelik II Referral Hospital.[6] The premises seemed to have served as the residence of an affluent family during the Imperial Era. The weathered main house was turned into the assembly hall, and former servants' quarters became offices. They added several temporary structures to serve as additional offices, stack rooms, and storage areas. The narrow corridor surrounded by those structures was occupied by several field vehicles, including an old four-wheel drive that seemed out of service. In sum, it demonstrated the features of a local welfare institution dedicated to outreach services.

There I interviewed Mengistu Zemene, one of the founding members and the head of the association. He was generous enough to share with me some of his personal stories. He had worked as a police officer in Addis Ababa under the previous regime. He lost his job in 1991 when the socialist government collapsed, and the Ethiopian Peoples' Revolutionary Democratic Front (EPRDF, the party that ruled Ethiopia until 2019), took control of the country. The following year, he found out that he was HIV positive. He was also diagnosed with tuberculosis, but antituberculosis drugs were hard to find. His doctor referred him to the Medical Missionaries of Mary, a religious institute of the Catholic Church that had been active in Ethiopia since the 1960s. They established the Counseling and Social Service Center (CSSC) in Addis Ababa in September 1992 to support people with HIV and AIDS orphans (children who lost their parents to AIDS). They offered Mengistu successive counseling sessions, medical treatment, and material support. "This led to my physical and mental recovery, and after that, I decided that I wanted to help other persons with HIV," said Mengistu. The CSSC employed him as a social worker and offered him training opportunities in Ethiopia and abroad.

In 1993, he spent a month in Uganda, visiting institutions engaged in HIV support activities. This experience led him to decide that he should establish a peer support group with his HIV-positive neighbors in Addis Ababa. Several individuals and AIDS orphans became the group's founding members, which they named Gabriel Mahber or Saint Gabriel's Society. *Mahber* is an Amharic term that refers to a popular form of fraternity association among lay followers of the Ethiopian Orthodox Church. Each *mahber* is associated with one of the Christian patron saints; Saint Gabriel was the patron saint for Mengistu's *mahber*. Although *mahber* activities are associated with religion, their primary function is often to foster close social relationships. Members of a *mahber*, usually close friends, neighbors, or members of an extended family, have monthly meetings in one of their homes, sharing a meal and engaging in conversations.

Mahber and other forms of association life represent an essential aspect of urban sociality in Ethiopia. Since Addis Ababa's foundation in the late nineteenth century, its inhabitants have formed and maintained numerous associations in the forms of *mahber*, *iddir* (burial associations), or *iqub* (credit associations) to avoid possible isolation and to provide social, moral, and economic support for one another (Alemayehu, 1968; Asfaw, 1958; Mekuria, 1973). Mengistu's Gabriel Mahber was a small group with no more than eight members. However, he wished to use this culturally familiar setting to create a space in which persons with HIV could provide moral support to one another. His *mahber* offered its members a safe refuge where they could reaffirm their care for one another's lives in a society where their illness was seen as a sign of impending unspeakable death.

The role of religion in the epidemic was ambivalent. The fear of AIDS and subsequent death was often expressed in terms of religious ideas such as evil and sin. Followers of the Ethiopian Orthodox Church, the country's religious majority, who accounted for more than 40 percent of its population, often related the disease to God's will. A report compiled by a US-based NGO recorded an Ethiopian woman saying, "This disease is the result of our sin and our distance from religion. If we didn't commit sin, this thing would have never come. Thus God will be merciful for us if we get closer to our religion. If we do good things and obey God's law, there will be no disease that has no cure" (Nyblade et al., 2003, p. 20). Such a remark seemed open to both stigmatization (by regarding AIDS as God's punishment) and comfort (by implying God's mercy and possible cure). The latter aspect provided significant spiritual support for the Orthodox followers with HIV, including members of the Gabriel Mahber.

While his *mahber* was a significant achievement, Mengistu was not entirely content with the intimate nature of the group and the religious comfort it sought. He believed that HIV-positive people in Ethiopia needed a broader initiative—something like that he observed during his visit to Uganda. The state of social movements for people with HIV in Uganda was quite different. In Uganda, TASO was established in 1987 by medical professionals. By the mid-1990s, they had been providing an array of services, including medical treatment, home-based care, patient and family counseling, training of caregivers, and material support and income-generating activities (WHO Global Programme on AIDS, 1995).

To achieve his ends, Mengistu proposed that they establish an institution of openly HIV-positive people. Gabriel Mahber was an informal gathering that did not involve any public activity, and Mengistu wanted the new institution to be legally registered and more open to the public. However, many of his colleagues considered the idea unacceptable. When the new institution was finally established in 1997, its official name—Mekdim Ethiopia National Association—did not mention HIV or AIDS. Only a handful of people volunteered to serve as its founding members. The association's name was derived from the Amharic word *mekdim*, which means "prologue." This word comes from the Amharic verb *keddeme*, which means "to go ahead." Unfortunately, in 1997, social circumstances in Addis Ababa did not allow most individuals with HIV to think about moving ahead.

Entoto

In September 2002, Meseret's brother, a medical doctor in Jimma, came to see her and her young son, who still lived in her late husband's household. For Meseret, her brother was one of the few individuals she could trust during those days of the crisis. She told him that she had decided to move to Gunchire, a small district town located between her village and Welkite, where she hoped to share her experience and knowledge of HIV with the townspeople. Her brother asked her if she had thought it through. She answered, "Yes, I have thought it through." They went to Gunchire and walked together into the district health office. The officers were "startled" at the idea of a young rural woman moving to town alone and speaking out about AIDS. They tried to stop her because they thought she decided to expose her HIV status to make a living. They even indicated that they could offer her financial support, which, in their perspective, could allow her to keep her HIV status private. However, they finally agreed to support her idea, and the next day, they drove into her village, loaded her belongings onto their pickup truck, and helped her move to Gunchire.

Meseret started to talk to people about HIV while visiting marketplaces, schools, churches, and mosques in the district, accompanying the district health officials. Some of her acquaintances attempted to dissuade her because they believed she would "ruin her life" by doing so. Some others spread rumors that Meseret pretended to be HIV positive and talked about AIDS for the sake of money. In those days, it was not unimaginable for some desperate individuals to identify falsely as HIV positive because people knew that some welfare institutions would provide them with material and financial support. However, at the same time, "They could not imagine a lively young woman could have the virus. People believed that a person with HIV looked like a monster," said Meseret. Such warnings and slanders did not deter her until one day, her son, who was still very young, heard people talking about her illness. He returned home and asked her what the AIDS they were talking about was. "This annoyed me very much," Meseret said, "and I left the town for the holy spring."

Followers of the Ethiopian Orthodox Church believe that holy springs have the power to cure illnesses (Guevara, 2006). Testimonies of the miraculous healing of the disease by holy water circulate in the local media (Berhanu, 2006). Kate Pfizenmaier (2015), who participated in a community health project in Gondar, a city in northern Ethiopia, noted that the tradition offered a unique opportunity to reach marginal populations for HIV education and care because "even people from the most remote villages travel to holy water sites."[7]

The pilgrimage to holy springs was facilitated by the extremely high level of stigma in Ethiopian towns and villages during those days. In a study conducted in 2006 at some holy springs in Addis Ababa, a female interviewee explained that her neighbors started to despise her when she lost her husband to AIDS. They even did not want to hang their clothes on the laundry rope they shared with her, and one of them prohibited their children from playing with hers, according to the woman, who added that she "didn't wish such a misery happened to people" (Berhanu, 2006, p. 49).

These discriminating behaviors reflected the widespread fear of "contamination" that the bearers of the virus were perceived to cause. Another form of extreme discrimination was refusing to share food or coffee for fear of contamination. In Ethiopia, sharing food with others is considered an essential human behavior, as demonstrated in an Amharic proverb: "one who eats alone dies alone."[8] Likewise, sharing coffee with neighbors is essential to communal life, and refusing to do so may be considered an ultimate form of denial. Meseret told me she knew a woman with HIV who refused to take HIV drugs, saying, "It's better to take coffee and die than to take drugs and live." By this remark, the woman implied that if she started taking HIV drugs, neighbors would recognize her illness and stop inviting her to share coffee.

Meseret was one of those Orthodox followers who sought safe refuge at the holy spring sites. When she left the town, she headed for the holy spring on Entoto Mountain, which lies on the northern margin of Addis Ababa. The spring is preserved by the priests of the Entoto Maryam (St. Mary's) Church. Thousands of people with HIV live in shacks on the mountainside. The shacks are owned by local proprietors and rented out to those seeking spiritual relief. Recalling the day she arrived, Meseret told me, "When I got there, no one spoke. But there were so many people who were like that," in one of our interviews in August 2015. This part of her narrative sounded particularly obscure because she avoided, perhaps unconsciously, using the terms HIV and AIDS when recalling her time in Entoto. What she meant was that she met many people in Entoto who were living with the virus, but no one spoke openly about their HIV status or experience with AIDS.

The people there avoided talking about the disease as if the silence was part of the cure they sought. Meseret described a woman who lived in the room next to her: "She did not tell me what she was like, although I told her," meaning that the woman did not tell Meseret about her HIV status while Meseret disclosed her own. However, the woman asked Meseret to go to the spring and bring some holy water for her. Meseret had no idea how to carry out the task. The woman explained that she should find the hall by the holy spring, enter it when they called for the members of the second association,[9] and she could receive the water in this hall. Meseret did not ask her what the "second association" was, but she went to the spring and waited for the call. "And then, when they called for the members of the second association, a crowd of people went in, went in, and went in, and the hall was pretty full," Meseret recalled. Inside the hall, she saw an older woman standing beside her. Meseret asked if all those people were "like that" (meaning HIV positive). "My child," answered the woman, "all these people are like that." The words struck Meseret. "Immediately, I was relieved. My anxiety, immediately, was gone, was soothed." She learned that she was not alone and felt connected to all the people in the hall that day.

In December 2016, I visited the Entoto Maryam Church with an Ethiopian driver. The sloped street that leads to the foot of the mountain goes through the Shiro Meda Market, the massive congregation of traditional cotton cloth retailers. Past the part of the street congested with shoppers, passenger vehicles, and unloading trucks, there stands the St. Peters Tuberculosis Hospital. Founded in 1961 by Emperor Haile Selassie I, it is the country's largest center for tuberculosis treatment and research institute. It is located at the northern edge of the town at an altitude of 2700 meters—the highest point in the highland city of Addis

Ababa. Past the hospital starts the steep winding road that climbs the hillside covered with eucalyptus trees. The road leads to the top of Entoto Mountain: at an altitude of 3200 meters, it offers a full view of Addis Ababa. In 1881, Menelik II, then a provincial *nigus* (king) of feudal Ethiopia, constructed his palace and military camp at this strategic site. Entoto Maryam is one of two eminent churches founded on the mountain before Menelik relocated his court to Addis Ababa, which became Ethiopia's capital after his enthronement as emperor in 1889 (Bahru, 2002).

We parked the car at a market corner next to the church and walked down a path through eucalyptus trees. We did not walk very far before finding a couple of structures in a quiet ravine. The guard was initially puzzled by the sight of a foreigner, but when I explained in Amharic the purpose of my visit, his mood changed, and he agreed that we could visit the place. (However, photographing was not allowed.) He confirmed that this was the holy water Meseret was talking about. It was close to noon, and I saw no one coming for the holy water. They came in the morning and were gone by that time, according to the guard. He also told me that the source of the holy water, access to which was restricted to the clergy, was further down the valley. They used a motor to pump the water from the source to allow all the visitors to take it home or bathe on-site. They had bathing rooms for men and women followers. The prayer hall, where Meseret was struck by the large crowd of people "like her," was also there. However, it was far from being the massive structure I had imagined from her story.

After staying at Entoto mountain for several months, Mesesret realized she was running out of money. Her brother had given her a considerable amount before she left the village. However, in Entoto, she paid the house owner a substantial amount for rent. One of her neighbors noticed her problem and suggested she visit a certain association called the Tesfa Goh Ethiopia Association (hereafter, Tesfa Goh, meaning "dawn of hope" in Amharic). Meseret traveled to Addis Ababa to visit the office, where she met a man named Dereje. He asked if he could help her, but she did not know what to say. He sat her down, told her that he was also "like her," and urged her to speak. She took out her medical certificate and gave it to him, still without saying a word. Reading the document, he kept quiet for a while and then left the room. Suddenly, she heard laughter from outside the room and thought they were making fun of her for some reason. Dereje returned, gave her some advice, and indicated that they could offer her a means of income. She promised to come back another day. However, she did not return there, being too annoyed by the apparent ridicule.

Connections

Dereje Alemayehu was working as a history teacher when he learned he was HIV positive. He quit his job and joined the community on Entoto Mountain before becoming a member of Mekdim. He then secured a paid position at Tesfa Goh. Established in 1998, Tesfa Goh was the second association of openly HIV-positive people founded in Ethiopia (after Mekdim). By the time he met Meseret, Dereje was one of the most active peer counselors for people with HIV in Addis Ababa (Interview with Dereje Alemayehu on January 1, 2015).

Meseret later learned why Dereje had laughed at her when they first met. The problem was with the diagnosis written on her medical certificate. It stated that she had *yezemenu beshita*, meaning the "modern disease" or the "disease of our time." Ethiopians gave HIV this name when it was still considered an unspeakable illness. It is understandable that Dereje and his colleagues, who were already the leading HIV advocates in the country, found it funny—if not outrageous—to find that term written on an official medical certificate.

As one of the pioneering associations of HIV-positive people in the country, Tesfa Goh attracted international support. During this time, the global movement of HIV-positive people was gaining momentum. International development aid agencies provided generous financial support for HIV interventions in Africa. Members of Tesfa Goh traveled the country to conduct advocacy campaigns and to set up regional branches. Dereje became the head of the association's Addis Ababa branch and traveled abroad to participate in conferences and capacity-building courses in Kenya, Uganda, Thailand, and elsewhere. Meseret heard Dereje say, while recalling the early days of their activities, "We came out when we were seen as *chiraq*. We crowed like birds." *Chiraq* is an Amharic word that refers to an imaginary monstrous creature that eats humans. That they crowed as birds refers to the fact that they not only spoke loudly about HIV but that their voices were heard at conferences and on radio and TV.

Meanwhile, Mekdim was also expanding its activities inside and outside Addis Ababa. While Tesfa Goh emerged as the leading advocacy institution, Mekdim focused on establishing an array of care and support programs. They founded counseling programs with HIV-positive individuals and their families, as well as those involving both Christian and Muslim leaders. They also developed home-based care activities, recruiting volunteers to visit AIDS patients and provide basic care and support. Moreover, Mekdim's clinic began providing ART for its clients free of charge in 2003, when antiretroviral drugs were virtually inaccessible to most Ethiopian citizens. They provided 104 people with ART before the government started providing it in 2005. By 2010, Mekdim's annual budget had become more significant than that of the CSSC, which had initially served as Mekdim's incubator (Brehony, 2010).

Shortly after her first encounter with the HIV activists in Addis Ababa, Meseret left Entoto and returned to Gunchire, a small town where she began to speak out about AIDS. Some individuals with HIV followed her example and began sharing their stories. Some patients' families were unhappy that she had "damaged their reputations" by encouraging their HIV-positive family members to speak of their status publicly. "Let them defend their reputation. But you cannot hide your AIDS and your pregnancy. People will notice," Meseret said, recalling those days. "That was a big change for me when I think about it. I am grateful that I cured some individuals' lives."

She used the Amharic verb *adane*, which I translated as "to cure." The term may be interpreted, depending on the context, either as to cure (of illness), save (from hazard), or redeem (from evil). Here, Meseret's remark is better explained as her having "cured" some people with HIV of their fear and humiliation rather than having literally "saved" their lives or "redeemed" their souls. During her second stay in Gunchire, she gained a reputation as the leading figure in the local fight against AIDS. People listened to her in marketplaces, school classrooms, churches, and mosques. She ate at local restaurants and often found the bill already settled by an anonymous supporter. Her house owner used to tell her that her rent for the month was already paid. This was one of the most hopeful moments in her complicated life in Gurage society.

In 2004, she was invited to a conference for people with HIV in the Southern Region. The meeting was supported by UNICEF and organized by the Tila Association of HIV-positive Women (hereafter, Tila), a gender-based association established in April 2002 in Awasa, the capital of the Southern Region. During the conference, she noticed that many participants from the other administrative zones represented local associations of people with HIV, whereas no such association existed in her zone at the time. Back in the Gurage Zone, Meseret and her colleagues established the Fana Association of HIV-positive People (hereafter, Fana), the first association of its kind in that area. The zonal administrative authorities helped them prepare the legal documents and assigned them a *kebele bet* (municipal housing) in Welkite to be used as the association's office. Meseret was one of the association's 20 founding members. When they voted to choose the association's chief representative, Meseret garnered the most votes. However, they agreed to elect a male member who was the "most educated of all" as their representative, and Meseret served as the treasurer.

She continued living in Gunchire, some 40 km away from Welkite, and visited the office occasionally. In May 2005, members of Tesfa Goh visited the Gurage Zone for an advocacy campaign. One of their destinations was Gunchire. They remembered Meseret and asked her to tour with them, to which she agreed. After they visited several places in the Gurage Zone, Tesfa Goh's members decided to invite Meseret to a national conference of people with HIV held the following month in Nazret, now renamed Adama. The city is one of the largest commercial centers in Ethiopia, located 100 km southeast of Addis Ababa along the highway to Djibouti.

The conference started on June 27, 2005, at the Rift Valley Hotel, which was considered the "fanciest" hotel in the city and was immensely popular as a venue for meetings organized by international and national aid organizations stationed in Addis Ababa. Upon arriving at the place, Meseret realized that she had landed in a place she had never seen before, and she asked herself, "Did I win the DV lottery?" The DV lottery stands for the Diversity Immigrant Visa Program, which allows thousands of Ethiopians to immigrate to the US every year. The hotel looked so modern to the young Meseret that she felt she had been teleported to the US. She was astonished by the room equipped with Western-style bathrooms and a huge TV. She was also surprised that the participants were served breakfast, coffee, lunch, and dinner without being asked to pay.

She remembered that Ugandans led the training course. "In the first session, we were asked to share where we were from and what we liked and disliked. I said I was from Gurage, I did not like lies, and I liked *kitfo*," said Meseret. *Kitfo* is a Gurage dish made from chopped raw beef and seasoned with a copious amount of butter and spices. In Gurage households, *kitfo* is served during the Meskal festival celebrations and for other special occasions. Her answer was translated into English so that the Ugandans could understand her. They appreciated her response and started to focus on her. During the sessions, they repeatedly referred to her as the "little girl from Gurage." Indeed, in the conference room, she was by far the smallest of all, explained Meseret.

The conference Meseret attended was officially called the "Leadership and Advocacy Training Course" for the leaders of Ethiopia's emerging HIV movement. It was jointly organized by TASO, the Ugandan NGO that pioneered care and support programs for people with HIV, and ActionAid Ethiopia, the Ethiopian division of ActionAid International based in Johannesburg, South Africa. This conference was not the first occasion where many of the emerging HIV associations in Ethiopia assembled and engaged in discussion. One of the early opportunities of this kind dates to 2000, when the Center for African Family Studies (CAFS), an NGO based in Nairobi, organized a workshop to facilitate communication among leaders of the HIV movement in Ethiopia. TASO, ActionAid, and CAFS were among the major supporters of HIV advocacy in Ethiopia at the time. Conferences and training courses offered by these organizations gave the leaders of HIV-positive groups in Ethiopia the opportunity to network, discuss their initiatives, and integrate into a larger movement. Representatives from the 18 leading associations in Addis Ababa and other major regional centers, such as Bahr Dar (capital of Amhara Region) and Awasa, created the first national network of associations of HIV-positive people in October 2004. The networking body was initially named the Association of Ethiopians Living with HIV/AIDS. It was later reorganized as the Network of Networks of HIV Positives in Ethiopia (NEP+). The network's membership increased very quickly. In 2013, it claimed to have 450 member associations with a total of 170,000 members (Network of Networks of HIV Positives in Ethiopia n.d., 2013).

Detachment

However, the rapid expansion of NEP+ and its connections with foreign civic organizations could have irritated the EPRDF, then the ruling party of Ethiopia. The party became particularly hostile against civic organizations after the general election in May 2005, when it lost a considerable number of parliamentary seats, including all 23 seats in Addis Ababa, to some new opposition groups. The party leadership identified then proliferating civic groups and their foreign donors as a vital constituency of the opposition.

The unusual redundancy in the "network of networks" structure of NEP+ is the result of an effort to accommodate the movement within the country's political milieu: It decided to adopt "ethnic federalism," one of the EPRDF's central ideologies, to its structure. In August 1995, the country adopted a new constitution introducing a federal system composed of nine regional states and two chartered cities (Addis Ababa and Dire Dawa).[10] Each regional state is supposed to represent an ethnic group, as in the case of the Oromo Region (representing the Oromo, the largest ethnic group) and Amara Region (representing the Amara, the second largest), although there are some exceptions, such as the Southern Region, where ethnic groups are represented at the lower level, as in the case of the Gurage Zone. Although Ethiopia's constitution provides that ethnic-based regional governments are highly autonomous political entities, in reality, they are tightly controlled by the ruling party (Nishi, 2005). NEP+ has adopted the federal system by fashioning itself as a national network representing regional networks of HIV associations.

In each region, the associations organize a regional assembly as their decision-making entity. Such a "decentralized" structure makes it easier for regional governments to closely monitor those associations' activities. Furthermore, the primary responsibility of NEP+ in this new structure was to engage in negotiations with the federal government rather than to work with its member associations.

The policy of NEP+ to work closely with the government ensured its further growth. In November 2008, NEP+ signed a grant agreement with the Global Fund to implement a project that cost 26.6 million USD by its completion in June 2015.[11] The Global Fund requires all recipient countries to set up a Country Coordination Mechanism (CCM) to govern the entire process. The Fund defines CCMs as "national committees that submit funding applications to the Global Fund and oversee grants on behalf of their countries." The CCM is supposed to include "representatives of all sectors" involved in HIV response, such as the government, civil society, and HIV-positive people. CCM's functions include coordination and development of the funding request, nomination of the principal recipient, and supervision of the implementation of approved grants (Global Fund, n.d.). As a member of the Ethiopian CCM and the implementing body of a Global Fund project, NEP+ secured its position within the country's antiretroviral regime.

Meanwhile, the political environment became even more hostile in 2009 when the Ethiopian government issued a proclamation that severely restricted the activities and resource access of all NGOs, including the associations of people with HIV (Federal Democratic Republic of Ethiopia, 2009). It stipulated that NGOs engaged in advocacy activities should not receive more than 10% of their revenues from foreign sources.[12] Furthermore, in 2011, the government issued a new guideline that effectively restricted financial transfers from foreign NGOs to national NGOs. The procedure demanded that all (foreign and national) NGOs operating in Ethiopia should not spend more than 30% of their budget on overhead costs. This might seem to be a reasonable regulation that would prevent some NGOs from paying unreasonably high salaries to their staff. However, very strangely, the guideline counted all financial transfers from foreign NGOs to national NGOs as administrative costs (Development Assistance Group, 2013). During those days, many notable foreign NGOs operating in Ethiopia shifted their priority from direct operations to assisting national NGOs' operations. As a result, a considerable part of those foreign NGOs' budget was financial transfers to national NGOs. The new regulations obliged foreign NGOs to choose between shifting back to direct operations or closing country offices. Simultaneously, the new rules significantly restricted access to foreign resources by national NGOs, including associations of HIV-positive people.

Consequently, the Global Fund and several other foreign agencies became the only sources of financial support for the associations of HIV-positive people, and the government became the sole channel through which those associations could access the resources. The Global Fund's CCM system, which was initially meant to secure an inclusive structure of the project governance, ironically helped NEP+ survive the hostile political milieu in Ethiopia. The Fund's project allowed NEP+ to provide financial support for its participating associations to conduct essential care and support activities, such as home-based care, income-generating activities, and nutritional support, for their members. Such access to this enormous fund cemented NEP+ as the focal point of the HIV movement in Ethiopia. In total, 104 associations, including Mekdim, Tesfa Goh, Tila, and Fana, participated in the implementation of the project. However, at the same time, the network of HIV-positive people's associations in Ethiopia was absorbed by the state-led antiretroviral regime. Furthermore, in the increasingly hostile political atmosphere, NEP+ and its member associations were detached from the global network of HIV movements financially and politically.

Notes

1. The official name of the region is the Southern Nations, Nationalities, and Peoples' Regional State.
2. HIV prevalence among adults (15–49 years old) peaked at 7.8% in 2000 and 2001 in the area containing 19 countries that UNAIDS categorized as the Eastern and Southern African Region, compared to the global figure of 0.7% in the same years.

3. The conference was titled "African Development Forum 2000 AIDS: The Greatest Leadership Challenge" and opened by K. Y. Amoako, then the Executive Secretary of the United Nations Economic Commission for Africa (ECA). It was held at Africa Hall, the main conference room at ECA's headquarters in Addis Ababa. I attended the conference as a member of the Japanese Embassy, where I served as one of the staff in charge of economic cooperation between 1999 and 2002.

4. Between 1998 and 2000, Ethiopia and Eritrea were engaged in a border conflict that is believed to have taken some 100,000 lives on both sides (International Crisis Group, 2003). Whether or to what extent the war affected the HIV epidemic in both countries is unclear. Statistical data from the Tigray Region (one of the Ethiopian regions bordering Eritrea) showed no evidence of a general increase in HIV prevalence in either civilian or military populations associated with the war (Berhe et al., 2005).

5. Another reason is that the revision in female rights was only effective in the Southern Region when the regional government proclaimed its version of the Revised Family Code in 2004. Widows' rights to property tended to be contingent on the provisions and applications of other legislations concerning land administration, which often favored men. Even after the regional family codes were enforced and land legislation was revised in favor of women, widows' access to land has often been denied or restricted in practice (Leckie et al., 2019; Tura, 2014).

6. Established in 1909, the Menelik II Referral Hospital is one of the first modern hospitals in the country.

7. However, since ART became widely available in Ethiopia, the use of holy water has been contested by health experts who identify it as a barrier against ART enrolment and retention (Kloos et al., 2013; Wubshet et al., 2013). Some followers of Ethiopian Orthodox resisted taking antiretrovirals, claiming that they would compromise the power of holy water, while others accepted both (Berhanu, 2010). Local policymakers, often Orthodox followers, preferred to address the issue by promoting ART and holy water rather than arguing against the latter. WHO Ethiopia Office worked with the Ministry of Health and the Ethiopian Orthodox Church to protect holy springs from being contaminated by bacteria that could cause acute watery diarrhea or other waterborne diseases (World Health Organization, 2017).

8. Many Ethiopian parents tell their children that it is a bad manner to fail to say "*inibla*," meaning "let's eat" or "let's share the dish," when they

start eating while others by their side are not. (However, it is very rare that the children actually have to share the dish because the others are supposed to decline the offer.)

9. Zena Berhanu reported that there was an association of people with HIV at Entoto, the aims of which included the distribution of donations from charitable organizations and individuals and organizing the burial when one of its members died (Berhanu, 2006). The "second association" mentioned here seems to be a group related to (or similar to) the association reported by Berhanu.

10. The number of regions later increased to 11 after establishing the Sidama Region in 2020 and the South West Region in 2021, both of which split from the Southern Region.

11. NEP+ was one of the principal recipients of the Global Fund's Round 7 Project in Ethiopia, titled "Ensuring Quality HIV/AIDS Services by Consolidating and Strengthening Existing HIV/AIDS Prevention, Treatment, Care and Support Programs" (Mookherji et al., 2015).

12. See Articles 2-2, 15-2, and 15-5 of the proclamation mentioned above.

References

Alemayehu, S. (1968). Eder in Abbis Ababa: A sociological study. *Ethiopia Observer, 12*(1), 8–18.

Asfaw, D. (1958). Ekub. *Ethnological Society Bulletin* (University College of Addis Ababa) (8), 63–76.

Bahru, Z. (2002). *A history of modern Ethiopia, 1855–1991* (2nd ed.). Addis Ababa University Press.

Berhanu, Z. (2006). *Care and support and people living with HIV and AIDS at Holy water: An assessment at four selected sites in Addis Ababa* (Master thesis submitted to the Graduate School of Social work). Addis Ababa University.

Berhanu, Z. (2010). Holy water as an intervention for HIV/AIDS in Ethiopia. *Journal of HIV/AIDS and Social Services, 9*(3), 240–260.

Berhe, T., Gemechu, H., & Waal, A. D. E. (2005). War and HIV prevalence: Evidence from Tigray, Ethiopia. *African Security Review, 14*(3), 107–114.

Birdthistle, I., et al. (2019). Recent levels and trends in HIV incidence rates among adolescent girls and young women in ten high-prevalence African countries: A systematic review and meta-analysis. *The Lancet Global Health, 7*(11), e1521–e1540.

Brehony, E. (2010). Report on review of MMM Counselling and Social Services Centre, Addis Ababa, Ethiopia. *Medical Missionaries of Mary*. http://www.mmmworldwide.org/images/stories/pdf2010/mmm_cssc_review_2010_adis_ababa.pdf. Accessed June 10, 2019.

Development Assistance Group. (2013). *Intermediary INGO operations and the 70/30 guideline*. Development Assistance Group Ethiopia.

Federal Democratic Republic of Ethiopia. (2000). Revised Family Code Proclamation No. 213/2000, Federal Negarit Gazetta Extra Ordinary Issue 6/1, 1–92.

Federal Democratic Republic of Ethiopia. (2009). Charities and Societies Proclamation No. 621/2009, Federal Negarit Gazeta 15/25, 4521–4567.

Geissler, P. W., & Prince, R. J. (2010). *The land is dying: Contingency, creativity and conflict in western Kenya*. Berghahn Books.

Guevara, M. W. (2006). Holy water cures even HIV. *The International Consortium of Investigative Journalists*. https://www.icij.org/investigations/divine-intervention/holy-water-cures-even-hiv/. Accessed February 6, 2022.

International Crisis Group. (2003). *Ethiopia and Eritrea: War or peace? Nairobi*. International Crisis Group.

Izumi, K. (2007). Gender-based violence and property grabbing in Africa: A denial of women's liberty and security. *Gender & Development, 15*(1), 11–23.

Kloos, H., et al. (2013). Traditional medicine and HIV/AIDS in Ethiopia: Herbal medicine and faith healing, A review. *Ethiopian Journal of Health Development, 27*(2), 141–155.

Leckie, J. R., et al. (2019). *Protecting land tenure security of women and girls in Ethiopia: Evidence from the land investment for transformation programme*. United Nations Economic Commission for Africa.

Mekuria, B. (1973). *Eder: Its roles in development and social change in Ethiopian urban centers* (4th-year senior essay, School of Social Work). Haile Sellassie I University.

Mookherji, S., Ski, S., & Huntington, D. (2015). Tracking Global Fund HIV/AIDS resources used for sexual and reproductive health service integration: Case study from Ethiopia. *Globalization and Health, 11*(1), 21.

Network of Networks of HIV Positives in Ethiopia. (2013). *Global Fund Round 7 Project implementation evaluation report*. Network of Networks of HIV Positives in Ethiopia.

Network of Networks of HIV Positives in Ethiopia. (n.d.). *Strategic plan 2008–2012.* Network of Networks of HIV Positives in Ethiopia.

Nishi, M. (2005). Making and unmaking of the nation-state and ethnicity in Modern Ethiopia: A study on the history of Silte people. *African Study Monographs,* (Suppl 29), 157–168.

Nyblade, L., et al. (2003). *Disentangling HIV and AIDS stigma in Ethiopia, Tanzania and Zambia.* International Center for Research on Women.

Pfizenmaier, K. (2015). HIV prevention and counseling at holy water sites in Ethiopia. *University of Washington Department of Global Health.* https://globalhealth.washington.edu/news/2015/09/28/hiv-prevention-and-counseling-holy-water-sites-ethiopia. Accessed April 2, 2022.

Ramjee, G., & Daniels, B. (2013). Women and HIV in sub-Saharan Africa. *AIDS Research and Therapy, 10*(1), 30.

Simpson, A. (2009). *Boys to men in the shadow of AIDS: Masculinities and HIV risk in Zambia.* Palgrave Macmillan.

Swaminathan, H., et al. (2008). *Women's property rights, HIV and AIDS, and domestic violence: Research findings from two districts in South Africa and Uganda.* HSRC Press.

Tura, H. (2014). *Women's right to and control over rural land in Ethiopia: The law and the practice.* Social Science Research Network.

WHO Global Programme on AIDS. (1995). *Taso Uganda: The inside story, participatory evaluation of HIV/AIDS counselling, medical and social services, 1993–1994.* World Health Organization.

World Health Organization. (2017). Reaching key populations to prevent the spread of disease in Ethiopia. *WHO Regional Office for Africa.* https://www.afro.who.int/news/reaching-key-populations-prevent-spread-disease-ethiopia. Accessed April 2, 2022.

Wubshet, M., et al. (2013). Death and seeking alternative therapy largely accounted for lost to follow-up of patients on ART in northwest Ethiopia: A community tracking survey. *PLoS One, 8*(3), e59197.

5

Life

Abstract This chapter addresses the realities of life and suffering under the universal ART regime in Ethiopia. It exposes the ambiguous exercise of "triage," that is, the exercise of latently yet systematically distinguishing between those who are "worthy" and those who are "unworthy" of laying claim to a meaningful life. In this regime, local health institutions simultaneously promote universal treatment by disseminating free antiretrovirals and triage by referring those with excessive suffering to Meseret's association. As a result, her group has borne an extraordinary concentration of such individuals. I regard this practice as an exercise in triage because, under the current ART regime in Ethiopia, Meseret's association has been systematically deprived of the resources required to address its members' needs. The emphasis on antiretroviral coverage and cost-effectiveness has resulted in a shrinking focus on social initiatives, such as providing home-based care by local associations like Meseret's. Moreover, the virtual ban on foreign aid to non-governmental organizations in Ethiopia has precluded the possibility of alternative funding. The ART regime in Ethiopia has ensured that those with excessive suffering are rendered invisible in the landscape of therapeutic citizenship in the country.

Keywords Defunding · Home-based care · Multiple health burdens · Triage

"If I were not apprehensive at that time, I could have reached someplace by now. Sometimes, it is better to be daring than to be apprehensive," Meseret said in one of our interviews in December 2014. She recalled her experience in 2005 when she was exposed to the quickly expanding national HIV movement. When I added to her words, half-jokingly, "You could be living in America by now," she promptly replied, "Yes, I could. Really." Meseret knew that, during the formative years of the HIV movement in her country, quite a few leaders found their way to the US.

By the time of her exposure to the national movement, the free ART program had been launched across the country, endowing survivors of the epidemic with another chance to live. They were also invited to an emerging form of therapeutic citizenship, which was very different from the local space of cultural refuge—the one Meseret encountered in Entoto Mountain—where patients silently recognized each other's suffering.

When Meseret made her debut on the national network, she attracted the attention of her colleagues partly because she was the youngest among them. When she was in the "Leadership and Advocacy Training Course" in Nazret, she quickly retreated to her room after the day's sessions were over and avoided socializing with the other participants. Although her encounter with the national network changed the course of her life, she kept a certain distance from the other leaders. Moreover, she returned to the Gurage Zone to work for Fana, the small association she established with some HIV-positive colleagues. Meseret's choice, which she described as a result of her hesitance rather than her conscience, kept her away from the larger network that could take her to America. Instead, she found herself closer to those whose suffering was increasingly invisible in the landscape of therapeutic citizenship in the country.

When appointed as Fana representative, she moved to Welkite, where she faced significant financial hardship. Although the government promised to cover the association staff's salaries, the amount was below the subsistence level and was not regularly paid. She quickly exhausted the small amount of money she had accumulated in Gunchire. She also had a difficult time finding affordable housing. Although landlords did not refuse to rent a room because of her HIV status, she had to move out of a room when its owner suddenly increased the rent without warning. Another owner asked her to leave because the room needed to be painted.[1] Her hardship continued until she married a man who had a stable income and his own house in Welkite.

This chapter highlights that Fana was increasingly deprived of already scarce resources to manage its weighty responsibilities. As the TasP-based interventions took hold in Ethiopia, Meseret's association witnessed a series of practices through which the humanitarian project was steadily swallowed into the sphere of institutionalized indifference.

Multiple Burdens

Welkite is primarily a commercial town that grew randomly on a hill along the highway that connects Addis Ababa and the southwestern part of Ethiopia. The town center is occupied by long-established inns that provide food, drinks, and accommodation for coach passengers and lorry drivers, as well as recently built commercial buildings that accommodate bank branches, offices, and retail shops. Welkite has also served as the capital of the Gurage Zone since 1995. At the southwestern edge of the town, the road starts that branches from the main highway and leads to the highland districts, which is home to millions of Gurage farmers, including Meseret's relatives. Further down the main highway is another corner, from which a dusty rural road leads to the resettlement sites inhabited mainly by non-Gurage farmers, including those from the Amara Region in northern Ethiopia. These sites were developed in the 1970s and 1980s as part of the government program intended to solve the recurrent famine (Woldemeskel, 1989). However, life in those sites is often harsh because they are located in a relatively low area heavily infected with malaria and trypanosomiasis, a parasite-borne disease that kills cattle.

Such a profile of Welkite and its backlands has contributed to the town's diverse population in terms of ethnic backgrounds and economic status. Welkite is home to Gurage businesspersons, shop owners, petty traders, and zonal officials of various ranks. It also hosts members of destitute peasant families of both Gurage and non-Gurage origins who wish to make a living out of odd jobs or depend on the townspeople's goodwill. The town also sees seasonal migrant workers of non-Gurage backgrounds who engage in day labor at agricultural farms or construction sites. I could not find data showing the HIV prevalence in Welkite. However, a recent survey estimated the prevalence in urban areas in Ethiopia's Southern Region at 1.8% (Ethiopian Public Health Institute, 2020). The same data revealed that the prevalence was significantly higher among women than men. Welkite's population was close to 28,900, of which 13,800 were female, according to the last census in 2007 (Central Statistical Agency, 2010). The town's population has presumably more than doubled since then.

Two major health institutions serve the population of Welkite: the Catholic hospital located in the suburb and the public health center in town. The former institution, where Meseret and her husband tested HIV positive years earlier, is officially named Attat Our Lady of Lourdes Catholic Primary Hospital. Locals refer to it as Attat Hospital. It started to operate in the 1960s as a mobile clinic carried on the backs of four rented mules and became one of the few modern hospitals in the Southern Region during the following decades (Lidman, 2016). The latter institution, Welkite Health Center, had been far less significant until recently, when it was expanded to become a modern health center with multiple service units. Both institutions have an HIV unit that provides HIV screening, counseling, and ART services. Both institutions work closely with Fana and refer some clients, particularly those with social, economic, or other problems, to the association.

In August 2013, Fana had 221 members, 153 of whom were women. They represented a relatively small portion of the HIV-positive population in the town. The striking characteristic of Fana members was that they were prone to multiple problems, including poverty, isolation, and comorbid conditions. Between August 2013 and March 2014, Meseret and I surveyed 50 members who had reported severe issues. Most respondents were either engaged in an unstable occupation (14 day laborers and seven petty traders) or unemployed (13 individuals). Among the 37 women respondents, many were single (four never married, ten divorced, and 15 widowed), making them socially and economically vulnerable. Forty-one of the respondents reported having one or more health problems other than HIV, and 35 said that these problems affected their life either severely or to a certain extent. Stated issues included tuberculosis, cancer, gynopathy, mental illness, and hearing loss. Despite being on ART for years and adhering to their medication regimes, they continued to face difficulties: 48 respondents reported that they had already started ART, and 41 said that they had taken their medicine as prescribed during the past week.[2]

The predominantly poor and isolated features of Fana's members do not represent the HIV-positive population in Ethiopia. Statistical data show that the prevalence of HIV is significantly higher among the rich than among the poor in the country's general population.[3] However, "it is only the poor who come out to join the Fana association," explained Meseret, while "the rich stay at home." According to Meseret, some individuals with HIV who live in town suffer more than those who live in rural areas, despite the general assumption that they are better off in cities where they gain better access to health and welfare facilities. "You may stay in the village as long as you have relatives who protect and take care of you," explained Meseret. "But some of our members come to town alone and live alone. They do not have anyone to depend on." Some individuals with HIV live in devastating situations, and Fana members exert considerable effort to help them. In some cases, these efforts resulted in improvements in the client's life, as illustrated by the case of Desalegn, who was middle aged and previously worked as a trader. (Desalegn is not his real name. All names of Fana members appearing hereafter are pseudonyms.)

When Desalegn tested positive, he fell into despair and indulged in excessive drinking. He subsequently lost his business and property, but none of his family members was willing to help him. He was living on the street when Fana members found him. They talked to some local benefactors, one of whom agreed to provide him shelter. In Ethiopia, it is not uncommon for a lay family to offer a destitute person a simple hut, often roofed with a plastic sheet and attached to the owner's house wall. He did not stop drinking, and his health deteriorated. When Fana staff took him to the health center, his CD4 count was critically low. They arranged for him to start ART. However, they were troubled when they saw him begging on the street, displaying a container of antiretroviral drugs, which he used as a tool to emphasize his misery to passers-by. In persuading him to stop begging, they enlisted him in a micro-credit program run by Save the Children Ethiopia, the famous international NGO's national branch. The program loaned him 3,000 ETB (approx. 200 USD at the time), with which he stocked bibles and crosses to sell on the streets near local churches. Subsequently, his health recovered, and his living conditions improved. Later, he started to live in the room of a woman who was also a Fana member. However, the woman did not seem very happy with him. Sometime later, for reasons not clear to Meseret, he left town before fully paying back the credit. However, he called the Fana office from time to time to inform them that he was okay, taking the drugs, and working. Although not everything went well with his case, Meseret was content that his life changed for the better. "When he called us, we never asked him to pay back," said Meseret. "We were happy that his life improved."

However, there were also cases in which Meseret's efforts failed altogether, as illustrated by the case of Tirunesh, a young woman whose life was complicated by HIV, poverty, and a lack of stable relationships. She was living in a village near Welkite with her HIV-positive mother. When her mother died of AIDS, she went to live with her grandmother. Their livelihood was so precarious that she had to seek help from other people. Tirunesh learned to survive by relying on people who would offer her shelter and support but often ended up finding herself in trouble. She stopped taking her antiretroviral drugs when she started living with one of her aunts. Meseret suspected that the aunt told Tirunesh to stop taking the drugs because she did not want her neighbors to discover the girl's HIV status. Although her aunt promised that she would take her to the holy spring, she never did. Moreover, Tirunesh had to stop schooling because she performed heavy manual labor at her aunt's home, such as pounding coffee beans with a wooden mortar and pestle. Her skin started to show signs of her deteriorating immune function.

Being aware of her problem, some of the local health center's staff helped her by giving her some money from their own pockets. However, they finally referred her to Fana and asked them to put her in a foster family. Meseret agreed to take care of Tirunesh at home because there was little hope that any other family would volunteer. "The first thing I did was to take her to Attat Hospital," said Meseret. When Tirunesh started the medication, she was troubled by what seemed to be the side effects. Meseret made sure that Tirunesh went to bed immediately after taking the antiretroviral drugs. Otherwise, Tirunesh could become excessively talkative—she started to tell a long story about the mean things that some individuals said or did to her. Tirunesh sometimes fainted after taking the drugs, so Meseret had to accompany her whenever she went to the bathroom, located outside the main house, to ensure her safety. This was a little embarrassing to both of them, considering that Tirunesh was almost 20 years old. However, such symptoms declined over time as she continued medication. Meseret also put her in primary school, where she started as a third-grade student. Considering her age, Tirunesh could have felt a little uncomfortable in the classroom. However, in those days, classrooms in Ethiopian primary schools commonly included overaged students whose schooling had been disrupted for various reasons.

It seemed to Meseret that they had established a good relationship until she noticed that her money was missing from the drawers. Both Meseret and Tirunesh were increasingly frustrated, and the latter stayed away more often. In search of a solution, Meseret talked to Tirunesh's grandmother. She was deeply saddened by what Meseret told her but could not help. By that time, Meseret had found out Tirunesh had several relatives, some of whom were more or less well-off. However, they were not willing to intervene. One of them declined Meseret's request to show up for a talk by saying he had "some other business." Tirunesh finally left Meseret. Later, when Meseret asked the case manager at Attat Hospital, who was in charge of following up with each patient on HIV medication, he said he did not believe that Tirunesh was regularly taking antiretroviral drugs.

Indifference

"Some people say that living with HIV is not a big deal. I think they are wrong. Once you are involved [in each person's problems], you see how complicated it is," said Meseret. For many members of her association, living with HIV is only one of life's many burdens. Against the scale and diversity of the problems they face, the association's human and financial resources were severely limited. As an organization, Fana was composed of an advisory board, office staff, and members. Its daily activities were managed by four paid office staff: a representative (Meseret), a program manager, a secretary, and an accountant. They also had volunteer workers who were mainly engaged in home-based care (HBC) activities.

Fana had been conducting HBC activities for its members for several years. It had 17 HBC volunteer workers and 158 clients in August 2013. The Global Fund provided financial support for such HBC activities in Ethiopia, which covered a small transport allowance for each HBC worker. Apart from this allowance, they were not paid a regular salary for their services. Workers visited each client once per week. They attended to the client's needs, including bathing, cleaning, and preparing food. They also monitored the client's living conditions, listened to their problems, gave advice, and, if necessary, referred the client to a health center. However, some of the clients experienced issues that were far too serious for the HBC workers, as illustrated by the cases of two clients who died in 2013.

Bezawit was one of Fana's HBC clients. She worked for a local farm, but after giving birth, she was diagnosed with cancer and bedridden. She lived with a male day laborer, and they rented a hut in a village outside Welkite because they could not afford a room in town. HBC workers had to walk for an hour to reach Bezawit's place. They helped her by washing her clothes and cleaning her room but could do little to mitigate the severe pain she experienced from her terminal cancer. She could not eat, and the idea of leaving her young child tormented her. When she died, Fana members washed her body before her neighbors buried her.

Welde was another HBC client who was estranged from his family and lived alone. He could not earn money because he had tuberculosis and was chronically sick. He often moved in pursuit of affordable housing. When he became severely ill, Fana members tried to help him by collecting money. Because the association did not have the funds to pay for Welde's hospitalization, the members gave their own money and solicited donations from townspeople. When he died, Fana members sent for his family, but they denied ever knowing him. His body was taken by the municipal office and buried.

For Fana members, how deceased members were buried was often as crucial as how they were cared for while living. In Ethiopia, how a person is buried is often considered no less important than how they lived (see Chapter 7). Typically, family members of the dead person would take up the primary responsibilities of washing the body and conducting burial and mourning rites. In the absence of family members or relatives, as in Bezawit's and Welde's cases, neighbors are expected to take up the duty. Bezawit's neighbors failed to wash her body, knowing that it could mean an insult toward the deceased because of her HIV status and the intense odor that filled her hut. During her last days, she was bleeding continuously from her uterus. Because of the strong odor, few people, except her husband and the HBC volunteer, approached her. Meanwhile, Welde did not even have neighbors to bury him because he moved so often that few recognized him as a neighbor. Having a member not buried adequately by relatives and neighbors was lamentable, if not humiliating, for Fana members.

On March 8, 2014, Fana members were invited to a meeting with local health bureau officials. The members complained of some local health experts' differential attitudes toward HIV-positive patients, mentioning a case in which a critically ill patient was laid on a mattress along the health center's corridor. They requested that the government rent land plots for destitute members who could not afford to lease rooms from private owners. An increasing number of members were being forced to live in remote areas because of the sharp increase in room rent in town. This made it difficult for them to access health support, employment, and education.

One of the health officials responded by saying that he believed local health experts were doing well and that individuals with HIV should remember that the government was already supporting their lives by providing them with HIV treatment free of charge. He went on to say that individuals with HIV were expected to join the effort to protect society from HIV. He suggested that they could do so by countering the negative influences of "immoral" foreign films and opposing homosexuality, which was, according to him, "spreading" in Ethiopia. When the meeting was over, the discussion did not seem fruitful. Nevertheless, Meseret was not disappointed, probably because she knew, from her previous experience, the kind of answer they would get from the officials. What she had not expected was that her members were so vocal. "Did you think it is only me who speaks?" said Meseret to the officer. "My members started to speak, and [that's why] I was silent today."

In the same month, Meseret told me that she had heard a rumor that financial support for Fana's HBC activities would soon be terminated. It was in July 2014 that the financial support for Fana and other associations' HBC activities were suspended. One of my informants, whose name I do not cite here, told me that this decision was made based on the assumption that most individuals no longer needed HBC support. In the past, many AIDS patients were confined to bed and needed someone to come and take care of them; however, with the advent of widespread ART, fewer people are bedridden. This assumption was based on data showing a rapid decrease in AIDS-related deaths in Ethiopia. Simultaneously, such belief was a clear indication that such individuals as Bezawit and Welde, who suffered from serious health problems that could not be treated with antiretroviral drugs alone, were already made invisible in the landscape of therapeutic citizenship in Ethiopia.

What happened to the Fana association was similar to Kenneth Maes's study on the moral economy of community health volunteers in Addis Ababa (Maes, 2016; Maes & Kalofonos, 2013). HIV intervention programs conducted in the city often relied on local volunteers who were virtually unpaid for the care they provided to their clients. They were typically motivated by religious or fraternal moral values that upheld sacrifice and devotion. The donor agencies justified using unpaid volunteers by arguing that the unpaid workforce would ensure the sustainability of the health services even after the external funding was terminated. However, Maes argued that despite good intentions, using unpaid workers could not be sustainable and that many people remained vulnerable to poverty, unemployment, and poor health.[4] The health volunteers interviewed by Maes often felt powerless because they did not have adequate resources to address their clients' problems. Moreover, devoting time and labor to work that did not generate income to sustain the lives of their own families often became a source of frustration (Maes, 2016).

In August 2015, I held a group meeting with some members of the Fana association and asked them to share their opinions on the changes that antiretroviral drugs did or did not bring about. They admired the medicine for allowing so many people to get up and work for their survival. One thanked the late Prime Minister Meles Zenawi for making the drugs available.[5] "The drug's benefits are beyond description," expressed another. "We have to take them with love," maintained a woman who served as a Fana volunteer.

However, they also experienced continued hardships in life. "Are problems gone? They are there, on the people," said a female member, "Some take their medicines, coming home at night after spending the day carrying soil and stones.[6] To live, they have to work, eat, and swallow the medicines." Another member who worked as a *telaleki* (messenger) in town said that after walking around the whole day, she often felt a burning sensation in her legs and wondered if it was a side effect of the drugs. The female volunteer said that each of her clients faced different kinds of difficulties and that it was difficult to listen to them. "Some can endure it; others not," she posited.

Care and Alienation

This chapter has presented some aspects of the multiple actions and engagements that formed and transformed HIV care in Ethiopia. Some emerged in the form of a praying multitude in the sacred mountain or a small self-help group in the quarter of Addis Ababa, each offering refuge to and forging a sense of solidarity among sufferers. Moreover, as ART scale-up took hold in Ethiopia, survivors of the epidemic, including Meseret, were invited to an emerging form of therapeutic citizenship,

disposed to replace the silent community of sufferers. This process significantly altered the course of Meseret's life. Among the many despairing
women who lost their husbands to AIDS, she reestablished her life as a
young leader in the network of HIV movements.

Meseret's story is reminiscent of similar local responses in other
African societies, which had predominantly (though not exclusively)
female faces. HBC activities organized by TASO in Uganda are among
the first documented activities to respond to the suffering of those
affected by HIV. However, John Iliffe (2006) noted that the precise
origins of such initiatives are hard to locate and probably diverse. By
the end of the 1990s, HBC activities were well established in the
Southern part of Africa, including Malawi, Namibia, South Africa,
Zambia, Zimbabwe, and in the East Kenya, Tanzania, and, of course,
Uganda.[7]

In Mozambique, a Christian organization named Mufudzi initiated
an HBC program in the mid-1990s with a "pragmatic, palliative, and
spiritual orientation toward relieving suffering and addressing material deprivation" (Kalofonos, 2021, p. 132). Ippolytos Kalofonos, who
conducted ethnographic research with Mufudzi's volunteers, noted that
the organization recruited "motherly" women who were considered
"wise, capable of maintaining confidentiality, and gifted at confronting
suffering and death gracefully and compassionately" (2021, p. 133).

However, Arminda, one of the Mufudzi volunteers, felt abandoned
when the practice of HBC was profoundly altered during the following
years by de-emphasizing attention to basic needs such as food and
narrowly defining care as technically oriented work. Moreover, the labor
of the volunteers was exploited as a resource, made freely available by
widespread underemployment, to be extracted in the name of cost-
effectiveness (Kalofonos, 2021, pp. 134–135). In Ethiopia, too, Meseret
found her association increasingly marginalized and defunded within the
"pharmaceutically-centered model of public health," which emerged as a
byproduct of the ART scale-up (Biehl, 2007, p. 1119). Her story (as well
as Arminda's) represents another case in which "the focus on keeping
bodies alive with medicines may leave persons more vulnerable" when

pharmaceutical interventions "fill in for a politics that can address socioe-conomic inequalities and pursue a political program of change" (Prince, 2012, p. 549).

Surviving the HIV epidemic takes a mesh of care practices driven by needs and filled with meanings. Biomedical care, while vital, may not replace them or dominate the entire system. Instead, pharmaceuticals may become essential threads woven into the mesh of practices. Thus, my argument does not suggest that contemporary technologies for controlling a pandemic plainly and inevitably undermine the local forms of care and meanings of life they convey. How HIV care is conducted is "a function of the particular material–social assemblages in which care is enacted," despite the virtual singularity of the mainstream intervention framework (Rhodes et al., 2019, p. 323). The experiences of surviving the HIV epidemic in Ethiopia, particularly the stories regarding the ways individuals like Meseret sought cures for their excessive suffering and how they were eventually caught up in the sphere of institutionalized indifference, provide a vital premise upon which claims for meaningful care are made, and ways to otherwise develop healthcare actions and engagements are sought.

Notes

1. Meseret explained that discrimination against people with HIV did not diminish, but indirect forms of discrimination replaced more direct ones as a result of HIV education. A study based on the Ethiopian Demographic and Health Survey in 2016 reported that 59.2% and 35.65% of sexually active Ethiopians had a moderate and high level of stigma against HIV, respectively. Stigma in this study included various forms of fear and prejudice that were perceived or experienced by the respondents. For example, fear of social judgment was there when the respondents reported that they would feel ashamed if someone in the family had HIV (Feyasa et al., 2022).

2. I believe the self-reported adherence in my survey more or less reflects the actual medicine intake, although some respondents might have exaggerated adherence. It is difficult to determine to what extent the patients'

self-reports on medication adherence reflect reality. However, a meta-analysis revealed that self-reported adherence is significantly related to adherence as assessed by other indirect measures, such as electronic drug monitoring and pill count, despite possible biases involved in the self-reporting method (Simoni et al., 2006).

3. Systematic analyses indicate that HIV prevalence is higher among the wealthier portion of the Ethiopian population (Ejigu & Tadesse, 2018; Lakew et al., 2015), despite arguments that associate poverty with HIV risk in the country (Kloos & Mariam, 2000; Sori, 2012). These findings do not necessarily contradict each other because there are specific conditions in which deprivation leads to increased exposure to HIV and other health hazards. For example, Alene et al. (2019) found a higher prevalence of HIV and tuberculosis co-infection in less wealthy areas of the country.

4. As Maes noted, the Ethiopian government moved against the use of unpaid community health workers in the first decade of the twenty-first century when it started to deploy health extension workers (HEWs) who were paid government employees. While HEWs contributed to improving the health of the population, their function was virtually limited to conveying a predetermined health information package (Bilal, 2012). As a result, responding to the care needs of individual patients remained with unpaid volunteers (Maes, 2016).

5. Meles's contribution to the ART expansion in Ethiopia is unclear to me, though clearly, it could not have come to fruition without his endorsement. His name was mentioned in the group discussion probably because they remembered him as a charismatic leader, and the later years of his service as the prime minister (1995–2012) overlapped the rapid expansion of ART in Ethiopia.

6. They worked as day laborers at construction sites to make a living.

7. On the other hand, West Africa was remarkably slow to organize support for home carers, even taking into account that the epidemic spread more slowly there and prevalence was generally lower (Iliffe, 2006, p. 108).

References

Alene, K. A., et al. (2019). Spatial patterns of tuberculosis and HIV co-infection in Ethiopia. *PLoS One, 14*(12), e0226127.

Biehl, J. (2007). Pharmaceuticalization: AIDS treatment and global health politics. *Anthropological Quarterly, 80*(4), 1083–1126.

Bilal, N. (2012). Health extension program: An innovative solution to public health challenges of Ethiopia, a case study. *USAID Health Finance and Governance Project*. https://www.hfgproject.org/wp-content/uploads/2015/02/Health-Extension-Program-An-Innovative-Solution-to-Public-Health-Challenges-of-Ethiopia-A-Case-Study.pdf. Accessed April 20, 2020.

Central Statistical Agency. (2010). *Population and housing census of 2007: Report for southern nations, nationalities and peoples' region, part 1: Population size and characteristics*. Central Statistical Agency, Federal Democratic Republic of Ethiopia.

Ejigu, Y., & Tadesse, B. (2018). HIV testing during pregnancy for prevention of mother-to-child transmission of HIV in Ethiopia. *PLoS One, 13*(8), e0201886.

Ethiopian Public Health Institute. (2020). *Ethiopia population-based HIV impact assessment 2017–2018: Final report*. Ethiopian Public Health Institute.

Feyasa, M. B., Gebre, M. N., & Dadi, T. K. (2022). Levels of HIV/AIDS stigma and associated factors among sexually active Ethiopians: Analysis of 2016 Ethiopian demographic and health survey data. *BMC Public Health, 22*(1), 1080.

Iliffe, J. (2006). *The African AIDS epidemic: A history*. James Currey.

Kalofonos, I. (2021). *All I eat is medicine: Going hungry in Mozambique's AIDS economy*. University of California Press.

Kloos, H., & Mariam, D. H. (2000). HIV/AIDS in Ethiopia: An overview. *Northeast African Studies, 7*(1), 13–40.

Lakew, Y., Benedict, S., & Haile, D. (2015). Social determinants of HIV infection, hotspot areas and subpopulation groups in Ethiopia: Evidence from the National Demographic and Health Survey in 2011. *British Medical Journal Open, 5*(11), e008669.

Lidman, M. (2016). *In Ethiopia, a little round house is the strongest medicine against women's labor complications* (Global Sisters Report). https://www.globalsistersreport.org/news/ministry/ethiopia-little-round-house-strongest-medicine-against-fistula-40896. Accessed on March 13, 2021.

Maes, K., & Kalofonos, I. (2013). Becoming and remaining community health workers: Perspectives from Ethiopia and Mozambique. *Social Science & Medicine, 87*, 52–59.

Maes, K. (2016). *The lives of community health workers: Local labor and global health in urban Ethiopia.* Routledge.

Prince, R. (2012). HIV and the moral economy of survival in an East African City. *Medical Anthropology Quarterly, 26*(4), 534–556.

Rhodes, T., et al. (2019). The social life of HIV care: On the making of 'care beyond the virus.' *BioSocieties, 14*(3), 321–344.

Simoni, J. M., et al. (2006). Self-report measures of antiretroviral therapy adherence: A review with recommendations for HIV research and clinical management. *AIDS and Behavior, 10*(3), 227–245.

Sori, A. T. (2012). Poverty, sexual experience and HIV vulnerability risks: Evidence from Addis Ababa, Ethiopia. *Journal of Biosocial Science, 44*(6), 677–701.

Woldemeskel, G. (1989). The consequences of resettlement in Ethiopia. *African Affairs, 88*(352), 359–374.

6

Ajyet and *Jegna*

Abstract This chapter examines various facets of Meseret's life as embedded in, but not defined by, local moral perspectives, particularly regarding the problems of care and womanhood. In the local social milieu, Meseret's involvement, on the one hand, is often understood in terms of a heroic "fight against AIDS," framing her life as comparable to that of a *jegna* or war hero who protects the community from foreign enemies. On the other hand, Meseret's own account of her life suggests that her moral subjectivity revolves around the image of an *ajyet* or caring woman. To better understand Meseret's moral endeavors in contemporary Ethiopian society, I look into local interpretations and reinterpretations of gender relationships. I also examine local efforts to combat the HIV epidemic, many of which are led by traditional male leaders and driven mainly by the idea of defending rural women from the virus.

Keywords Caring woman · Gender relationships · Local HIV responses · Womanhood

I wrote in Chapter 4 about how I first met Meseret in September 2007. My second visit to Meseret, in February 2008, unfolded as follows: On arriving at Welkite, I called her mobile phone number, hoping that she had not forgotten my name. She remembered me only after we ended the first call and called me back. During this visit, she showed me a local magazine that contained a short article about her life.[1] It explained that Meseret was living in a rural village when she married a soldier in 1995. He went to war in 1998 when the border conflict between Ethiopia and Eritrea erupted. She contracted HIV from her husband, who died shortly after returning from the war. The article explained, "Her aspiration was to stay in the village, follow the disciplines and the culture of the villagers, and live as an *ajyet*—it would have been her choice if she could have made it."

Ajyet is a Gurage term used to refer to a skillful and respected wife who keeps her household neat and takes good care of her family and, above all, her husband's guests. Here, the anonymous writer presents Meseret as a martyr of two significant ordeals that simultaneously assailed the nation: The war against Eritrea and the HIV epidemic. The virus destroyed her aspiration to become an *ajyet*. However, such an assumption serves also as a premise to place her as the invincible *jegna* or "brave one" fighting against HIV. "Thanks to AIDS, I have reached where I am now. Otherwise, I was an ordinary rural woman," said Meseret in our first interview (see Chapter 4). This remark appeared to confirm her transformation from an unsuccessful *ajyet* to an unyielding *jegna*.

As our conversations continued over the years, I started to question the story of sheer transformation. I became more concerned about what had been continuous in Meseret's life trajectory and moral engagement. I became more convinced that her consistent inclination to be involved in other people's life troubles and ability to see "how complicated they are," as elaborated in the previous chapter, is not the outcome of her transformation from an *ajyet* to a *jegna*. However, this is not to say that her initial aspiration to become an *ajyet* or follow the cultural archetype of a caring woman continued to drive her moral engagement.

The rest of this book is dedicated to exploring how the lives of Meseret and some other Ethiopian women were shaped by their encounters with disruptive yet formative events, including but not limited to the

HIV epidemic in Ethiopia. This journey is navigated by the following questions Biehl (2007, p. 8) asked:

> Who empirically is the agent of this making and remaking of culture? How is this process mediated by individual lives? What do psychological structures and modes of experience contribute to the work of culture? And how do modes of subjectivity intertwine with particular configurations of political, economic, and medical institutions? In other words, how, under quite new conditions, do people value life and relationships and "enact the possibilities they envision" for themselves and for others?

I elaborated in the previous chapters on how Meseret's life trajectory intertwined with some "configurations of political, economic, and medical institutions" in contemporary Ethiopian society. In what follows, I will trace the entanglement with more focus on the cultural context in which lives unfold in the shadows of the HIV epidemic, gendered cultural norms, and patriarchal social order. Overlapping stories from several women's lives will be presented. These stories include bold escapes from patriarchal social fabrics, endeavors to maintain caring relationships amid differential systems, memories of defiance and affection shared through maternal lines, and alternative sociality imagined in life at the margin. They amount to what I refer to as the "culture of defiance" that underpins Meseret's actions and engagements which, she claims, have cured the lives of others.

This chapter elaborates on the gendered contexts within the Gurage society during and before the HIV epidemic. First, it explains how people interpreted Meseret's efforts within the context of the gendered norms of Gurage society. Then it examines how local male leaders responded to the epidemic to conclude that their responses were articulated in conformity with existing gender norms and reinforced patriarchal social order. Finally, I focus on the female agency to problematize gender boundaries and imagine the other forms of sociality.

The Fight Against AIDS

Meseret claims to be the first person in the Gurage Zone to make her HIV-positive status known to the public. She did so when she was 18 years old and lived in Gunchire, a rural town in the Gurage Zone. Antiretroviral medicine had not yet been available, and fear and stigmatization were prevalent. "They did not believe that I was HIV positive. They looked at me and said, 'Such a pretty young woman cannot be an AIDS patient.' People had a horrible image of AIDS at that time," explained Meseret. She worked with the district health office, visiting schools, government offices, and marketplaces to talk about HIV/AIDS. She told those who did not know their HIV status not to be afraid of HIV testing. She advised those who were depressed about their HIV-positive status to stop hating themselves. When recalling her days in that town, she always added, "You don't know how much they loved me. When I ate at a local restaurant, the waiter often told me that my bill had already been paid. Even my house rent was paid by an anonymous person every month." People increasingly perceived her as a protector in their fight against AIDS.

In September 2008, Meseret, her son, and I visited her home village in the Enemorna-Ener District, located 70 km from Welkite. Our visit occurred during Meskal. Based on a story of the rediscovery of the lost cross used to crucify Jesus Christ, Meskal is one of the most important holidays celebrated by followers of the Ethiopian Orthodox Church. For the Gurage Orthodox followers, Meskal has a particular cultural significance as the occasion for the annual family reunion. (For the Muslim Gurage, the Arafah serves as such occasion.) To celebrate Meskal, Gurage families in Addis Ababa and other cities head for their home villages by coaches or Toyota passenger vans—the latter usually serve as means of public transport in town. The last part of the trip often involves a harrowing drive on muddy terrain, as it happens during the end of September, which marks the end of the major rainy season in the Ethiopian highland.

Meskal is also the most popular season for Gurage weddings. We attended several wedding ceremonies in the vicinity. At every household we visited, women offered us an unlimited amount of *kitfo*, minced raw beef marinated in seasoned butter, which is essential for Gurage festivals. An Amharic saying goes, "the face, rather than the feast [explains hospitality]." Every woman was convinced that the most crucial element of serving *kitfo*, particularly during such a festival, was demonstrating generosity. I was given a piece of thick green leaf, which was folded to an inverse cone to hold out to the woman who would serve the *kitfo* in it, along with a hearty portion of hot butter poured on top of the meat. I noted that minced raw beef's ability to absorb butter was surprisingly high. I begged her to stop, but she continued until she was convinced that a sufficient amount of butter was served. When I came close to finishing the first portion, I was offered another. Similar interactions were repeated several times a day in every household we stepped in.

As we toured the houscholds of Meseret's near and far relatives, I heard a man say of her, "If she were a man, people would have counted her as a *jegna*." The man said the words to praise what she had done to protect people from HIV. Although I was aware that a *jegna* is masculine in its original sense, I asked him if a woman could be a *jegna*. He quickly answered, "Well, she is the first female *jegna*," hoping to dispel the male-centric tone of his earlier statement. Meseret responded only by laughing.

Jegna is an Amharic term referring to a brave man and is heavily loaded with local gender values: a *jegna* is supposed to embody manly virtues, such as *gobez* (clever and robust) and *defar* (fearless and brave). A contemporary Amharic dictionary defines a *jegna* as one who sacrifices and carries out good deeds for one's country or fellows (Yeltyopya Kwankwawoch Tinat na Mirmir Maikel, 2001). Thus, to identify Meseret as a *jegna* is to consider her as someone who sacrificed her life to protect fellow Gurage in the fight against HIV.

I wonder to what extent such an estimation explains Meseret's experience of surviving the epidemic. "If I were not apprehensive at that time, I could have reached someplace by now. Sometimes, it is better to be daring than to be apprehensive," said Meseret, recalling her early days in the national HIV movement (see Chapter 4). She indicated that she could have gone to America only if she had been daring enough when exposed to the movement in 2004. Then again, it may not be sheer lack of courage that brought her back to and kept her in the provincial town of Welkite. She did not lack the courage to face an unfamiliar world when she left her village to be the first one to make her HIV-positive status open in the Gurage Zone when people perceived AIDS as a cursed illness. Lack of conviction, rather than courage, would better explain her hesitation to follow the ladder that could have "taken her to America."

The story of transformation from an unsuccessful *ajyet* to an unyielding *jegna* does not explain her continuous engagement with people living with multiple health burdens and life complications. Despite her departure from an "ordinary rural woman," Meseret emotionally joined the rural community in mourning the loss of a caring woman, as described below.

The Caring Woman

In August 2012, a woman named Murgad Werku died at 67 years old in a rural Gurage village. She was married to a wealthy man and had eight children. Her husband was a successful businessman in Addis Ababa, and many of their sons were engaged in the family business. While her husband and grownup sons spent most of their time in the city, Murgad stayed in the village to supervise her husband's land. Such a spatially extended family life has been common among the Gurage, particularly during the twentieth century when Gurage men migrated to towns and established their reputation as brisk businesspeople.

As Meseret had lived in a nearby village during childhood, she and Murgad had known each other for many years. Meseret and I attended her funeral. When we reached the household Murgad used to live in, we saw several buses parked on the road in front of the gate, indicating that a number of people from Addis Ababa had already arrived there. They hired buses to reach the village in time for her burial. Her coffin was laid in a guest hall built in a large compound. The room was filled with low, intense waves of groans each time a crowd of women and men poured into the room as they arrived from nearby settlements. At times, some women fell on the floor to demonstrate their acute grief.

While the crowds continued to arrive and repeatedly filled the room with their wailing, a couple of men were busy on the other side of the compound boiling a large amount of beans in two iron pots, each of which was the size of an oil drum. They were preparing *nifro* (boiled beans) to serve the guests. As the time for the departure of her coffin approached, village elders on horseback gathered to escort it to the church. While waiting for the coffin to come out, the horsemen marched in front of the gate, back and forth in a row. Soon, a number of women joined the march, crying loudly, raising and waving their hands as a sign of their deep grief. Murgad was buried in the cemetery at the Orthodox Church in a nearby settlement. After the burial, the mourners sat silently on the benches lined up in the compound and had coffee and *nifro* before going home.

The day after the funeral, Meseret explained how she admired Murgad Werku and regretted her death.

> I have a lot to tell you about her. She received every guest in the proper manner. She never made differences in the way she treated people, whether important or ordinary or rich or poor. She shared dishes and coffee cups with artisans (who are often considered outcasts in rural Ethiopia). She treated people equally and thought well of everyone. If she learned you were in distress, she gave the right advice, saying, "My son, that is not the way you think." She offered to those who did not have. If she learned of someone who lost their cattle, she gave away one of her female calves. You could see her cattle in many households in the village. Many children, particularly girls who had lost their parents, used to meet at her household to celebrate the Meskal. You cannot find such a woman today. She was our mother. We have lost our valued mother. She did not live an easy life. She experienced hardship in her younger days. She had reached a point at which she was tolerating all the challenges in her life. Then she suffered from a long illness in her last years. She deserved a longer life.

Murgad Werku lived an admirable life as a Gurage woman. She was a caring woman. Showing care and affection to a sufferer is considered an essential part of moral life in Gurage society. For instance, "may my womb open split" is an expression common among female Gurages to show concern for fellow women's well-being. It originated in the context of childbirth, whereby accompanying women say these words to comfort the woman in labor (Shack & Marcos, 1974, p. 83). Thus, the expression can be interpreted as "May your pain be mine."

Interestingly, this expression appears in a chant dedicated to a legendary war hero. A young maiden is the narrator of a 27-line chant recorded by Shack and Marcos (1974, pp. 81–84). She celebrates Nigis Amerga, a prominent Gurage chief who fought against neighboring ethnic groups, such as the Oromos, in the early twentieth century. In the latter part of the chant, she addresses Derwet, the hero's mother:

> Oh mother of the brave, may my womb open split!
> Oh mother of the great man, may my womb open split!
> Why be open split? For bearing
> all and only good men,
> all and only wise men

In these lines, the expression "may my womb open split" serves as a way to confirm male supremacy, overshadowing its original sense of female solidarity.

Likewise, one could praise Murgad Werku as the wife of a successful Gurage man and for giving birth to eight children who managed parts of the family business in Addis Ababa, highlighting the patriarchal aspect of local society. However, in her narrative, Meseret celebrated Murgad as the emotional mother of her neighbors, particularly those troubled.

The Premarital Screening Campaign

Gendered aspects of Gurage social norms inevitably manifest in some local responses against the HIV epidemic. During my fieldwork, I recorded two such responses promoted mainly by local elders: the premarital screening campaign and the "cultivate the backyard of your neighbor" campaign. As I elaborate below, these campaigns were articulated in conformity with the local gender norms and conducted in a way that reinforced patriarchal order within the Gurage society.

In 2007, the Ethiopian government launched the "Millennium AIDS Campaign" as a national mission to be pursued during the year when the country celebrated the Ethiopian Millennium or the transition to the third millennium in the Ethiopian calendar.[2] The campaign was specifically designed to increase HIV screening. As a result, the number of HIV tests conducted in the Gurage Zone rose sevenfold in two years— from 14,063 in 2006 to 99,177 in 2008. The number of individuals who tested HIV positive increased from 737 to 1736 during the same period.[3] This startled the Gurage elders, although the number reflected the increased test coverage rather than the rising incidence of infection.[4] The elders were afraid of the advent of the epidemic owing to their belief that their people were particularly susceptible to the virus because of their mobility.

It was common for young Gurage, particularly men, to migrate from rural areas to towns and cities in search of job opportunities. At least half a million Gurages live in Addis Ababa, the capital of Ethiopia. A typical life course for a young man was to migrate to an urban area, such as Addis Ababa, in his teens and engage in a trade or business run by immediate family members or other relatives. When he and his family felt that his economic life was well established, he would marry a rural woman. Gurage men born and raised in cities also often married rural women. Those wives were often asked to stay in the village to tend to their husbands' land. Urban Gurage men typically own plots of rural land inherited from their fathers.

However, during the HIV epidemic, marriage between urban men and rural women was perceived as presenting serious health risks for the rural community. Since 2003, rural elders have increasingly asked young couples to undergo HIV tests as a precondition for recognizing their marriage. Seyfu Welde, one of the notable Gurage elders and head of the local Ethiopian Red Cross branch office, shared in one of our interviews in 2010 that a young couple was asked to be tested twice. The first test was required immediately before the rite of engagement, which took place several months before marriage. The second was required before the marriage ceremony. Seyfu explained that the second test was necessary because of the "window period" between the first infection and the point at which the HIV test can reliably detect the infection. He also estimated that "at least 80%" of all young couples preparing to marry underwent premarital screening.

Ethiopia's family code provides that "marriage may be concluded before an officer of civil status" or "following the religion or custom of the future spouses" (Federal Democratic Republic of Ethiopia, 2000, p. 3). However, marriage following customs has been the most recognized form among the Gurage. If a Gurage couple wished to have their marriage fully recognized by the community, the preferred practice was to ask relevant elders to endorse the union.

In Gurage villages, elders play a central role in maintaining social order and enhancing communal discipline. The Gurage customary law (*qicha*) is maintained by elders who confirm, revise, and apply the codes by consensus. Meetings among elders are held regularly at the village level. A larger council of elders (*yejoka*) is held once every several years, bringing together all significant elders. The council discusses issues such as solutions to disputes between clans and revisions of important laws. Customs are maintained as unwritten laws. However, some essential provisions had been compiled and published in the 1990s by the Gurage People's Self-help Development Association (GPSDO), a non-government organization led mainly by urban Gurage male elites.[5] The leaders include professionals, such as lawyers, bankers, and successful merchants who financed the organization's activities. One of their goals is to promote Gurage culture and norms in the changing socio-political environment by working closely with rural elders. In 2007, the GPSDO published a revised version of the Gurage customary law to include a provision for premarital HIV screening: Article 4, Section 3, which provides that a young couple should be tested twice before marriage. The same article also provides that a married couple should be tested again after the partners have lived apart for some time. If one of the parties should refuse to get tested, that partner will be ordered to leave home (YeGurage Hizib Ras Ages Limat Dirigit, 2007, pp. 12–13).

This intervention, aimed at protecting rural women to ensure the reproduction of the Gurage population and society, had a somewhat controversial outcome. One of the health centers in the Gurage Zone reported that a total of 17,379 individuals underwent HIV screening, and 51 (of whom 27 were women) were found to be HIV positive during the three years before June 2010. Among those who tested positive, only three (one woman) were unmarried.[6] One might doubt the extent to which the tiny number of HIV-positive results among the unmarried reflected reality. A more likely scenario was that most young Gurage who wished to get married received HIV tests outside their village before they faced "official" premarital screening. They did so in the hope of avoiding possible disappointment if they were found to be HIV positive in the middle of wedding preparations. Some experts have argued that premarital and postmarital screening campaigns are not practical because of the easy availability of forged HIV test results in Ethiopia.[7]

Cultivate the Backyard of Your Neighbor

Another consequence of the HIV epidemic was the loss of the labor force, leading to poverty in some rural Gurage households. In August 2009, I visited Tigist (not her real name), accompanied by a local health extension worker. Tigist had three children, the oldest of whom was a 12-year-old boy. Her husband died in 2004, presumably from AIDS-related complications. After experiencing poor health following her husband's death, Tigist was tested and found HIV positive. Her youngest daughter, who was five years old at the time, also tested positive for the virus. Since the death of her husband, Tigist and her family had encountered difficulties maintaining their household economy. While Tigist's husband was alive, he was a typical migrant laborer, working in Addis Ababa most of the year; he sent money to Tigist, who continued to live in the village. With the remittance, Tigist hired local men to cultivate their backyard, which constituted a field adjacent to the family residence and planted primarily with the staple crop known as enset. The perennial plant is a member of the Musaceae (banana) family, reaching a height of over five meters at maturity. It is a staple or co-staple food for approximately 20 million people in the southcentral, south, and southwestern parts of Ethiopia, including the Gurage (Borrell et al., 2020; Sahle et al., 2021).

The Gurage system for enset cultivation is labor-intensive. Gurage men transplant enset several times before it is ready for consumption. A dense forest of dozens or even several hundreds of grown enset plants towering in rows in the backyard is the source of pride and represents the household's sufficiency. However, unless the enset is transplanted at the appropriate growth stage, it does not develop to its full potential, making it difficult to obtain sufficient starch to feed the household. With her husband's death, Tigist lost her source of cash income and, along with it, the ability to hire the necessary labor. Without appropriate transplantations, the yield of her already meager backyard continued to decline.

Tigist's experiences are typical of households affected by HIV in Gurage villages. The phrase "cultivate your neighbor's backyard" often came up in my interviews with village elders and health workers when I asked them about possible community measures to cope with the household effects of HIV. "Cultivating the backyard" of a Gurage household generally refers to the planting and transplanting of enset. "Cultivate your neighbor's backyard" in this particular context indicates the communal responsibility to provide agricultural labor for households affected by HIV. People have mixed opinions about this idea. Some regard the course as a manifestation of the Gurage sociality. However, others are skeptical that it would be a temporary solution at best on the grounds that few communities would continue to bear such a burden.

The responsibility is supposed to materialize through *gyez* or labor cooperatives. In Gurage villages, several neighboring households form a group and perform farm activities in turn for each household. Separate *gyez* are formed based on gender, with the men's *gyez* being responsible for planting and transplanting enset and women's *gyez* in charge of processing enset for consumption. Men's *gyez* generally include five or six households, and membership is likely to remain stable for years. The members are not necessarily relatives. A *gyez* can be formed as the members see fit, based on criteria such as friendship or accordance with farming practices. *Gyez* members do not merely work together; they become, in a certain sense, economically responsible for one another's households. For example, a Gurage farmer explained that if one of the other members of his *gyez* died and left behind his wife and young children, his *gyez* would cultivate the deceased member's backyard without a reward until one of the sons was old enough to take part in *gyez* labor. It is also expected that a *gyez* member "lends" his *gyez* to another impoverished household, usually a relative who is not a member of the *gyez*. In

other words, he can ask the *gyez* to cultivate another household's backyard instead of his own. In Gurage farm villages, *gyez* are expected to
serve as a "safety net" to secure labor access.

However, the principle of mutual help did not appear to apply to
Tigist's household. Since her husband had relocated his economic base to
Addis Ababa, and because the money he remitted was used to hire labor,
they had not developed long-term relationships and a sense of mutual
commitment with other households through a *gyez*. Further, there did
not appear to be any relatives living nearby who could "lend" their *gyez*
to her.

In March 2010, I visited three households affected by HIV, one of
which was Arganesh's (not her real name). She was a mother in her
twenties with three children, of whom the oldest was an 11-year-old
boy. Arganesh's backyard was small and consisted of what local farmers
call "dry land," or soil unsuitable for enset cultivation. As in Tigist's
case, it was evident that her household was not wealthy even before her
husband's death. However, I was told that Arganesh's neighbors regularly
helped take care of her household in various ways. They assisted with the
enset cultivation in her backyard or took care of her livestock when she
had to go out of town to receive antiretroviral drugs. The properly tended
state of her small backyard was a clear indication of the contribution of
her neighbors. Shortly after we were invited to her house, neighbors gathered carrying large *jebena* (coffee pots) and *kolo* (roasted barley). In rural
Gurage, neighbors are responsible for assisting the household in hosting
a guest.

Arganesh's husband died three years earlier. When her husband's
health began to deteriorate, they suspected an HIV infection and visited
the health center together to get tested. The results showed that both of
them were infected. Her husband was devastated by the results, but she
accepted them calmly. According to the story told by the health extension
worker who accompanied me, Arganesh's neighbors were "grateful" that
she had announced her HIV-positive status immediately after receiving
the results of the HIV test.

It is usual for Gurage women to remarry after losing a husband, either
through death or separation. It was likely that the still-young Arganesh
would receive marriage proposals, or there could be a man who would try

to "force her" to marry him, even if she was not interested. Meanwhile, Arganesh had enough reason not to reveal her status since she could end up in poverty; announcing her HIV-positive status could narrow her chances of remarrying. Then, she would be left without labor to cultivate her backyard, given that her son was not yet old enough to take on that responsibility. Arganesh's proactive announcement of her HIV-positive status, despite the potential negative consequences, was seen as an act of concern for her neighbors' health. For this reason, the villagers decided to support her household as a sign of gratitude for her consideration.

I also visited Tigist's household, which I had seen the previous year. Her backyard improved a little, thanks to the help provided by neighboring farmers in transplanting the enset. However, it was unclear whether this help was a one-off event or something that could be relied on more continuously. I asked her if she was receiving sufficient support from her neighbors, to which she answered yes. However, shortly after I left her home, one of the neighbors' children, who had watched me talking to her, came running to tell me that she was not receiving adequate help, hoping that I would return to offer some material assistance.

The third woman was even more impoverished. After losing her husband to AIDS, she could not maintain her backyard, and she had to move into her parents' household, which also was in distress. Neither this woman nor Tigist appeared to receive appropriate HIV treatment. Although HIV treatment was provided free of charge, they did not have the money to pay the bus fare needed to visit the health center in a town more than 40 km away. During my visit to these two households, neighbors did not show up with coffee pots, indicating their isolated status.

The two campaigns described above are examples of local HIV responses articulated to fit the local sociocultural fabric. I doubt the premarital screening campaign lived up to initial expectations. Its impact on reducing HIV incidence is questionable, as explained previously. However, I suspect that the campaign had the unintended effect of reinforcing some patriarchal aspects of Gurage society.[8] In other words, it reinforced the solidarity between rural and urban male elites and their authority to control the familial and reproductive systems of their rural community, although such an effect was not intended by those who promoted the idea.⁹

I wonder what could have led to a more consistent implementation of the "cultivate the backyard of your neighbors" campaign. It attempted to mobilize the local labor force to help households affected by HIV with limited success. Gurage elders hoped they could achieve their goals by mobilizing the indigenous mutual help associations organized in line with the gender division of farm labor. Men were encouraged to assist women who lost their husbands to AIDS. Among the three cases I briefly observed, Arganesh's was interesting because of the mutual trust between the HIV-positive woman and her HIV-negative neighbors. Despite the campaign's potential to promote moral and economic engagement, the overall result seemed mixed, at best. I imagine the campaign could be more consistent if all the *gyez* involved were compensated for their labor, perhaps in the form of cash supplied by governmental or non-governmental institutions.

Gendered Norms and Division of Labor

In the previous sections, I described how Gurage elders articulated their HIV responses to fit the community's gender norms. The premarital screening campaign was an attempt to protect rural households from HIV by reinforcing the patriarchal social order. "Cultivate the backyard of your neighbors" campaign intervened in the domain of subsistence labor, which is divided along gender lines. It aimed to mobilize male labor power to support households affected by HIV.

Unlike the two campaigns led by male leaders, Meseret's moral engagement does not simply conform to the existing gender norms of the community. Her actions problematize rather than reinforce such presuppositions. I doubt that her exposure to the national HIV movement (Chapter 4) enhanced her tendency to do so, given that she kept some distance from the movement's leaders. Instead, I am inclined to explain her actions as informed by the "culture of defiance" that is prevalent, albeit not always obvious, among Ethiopian women. Meseret claims to be the first person in the Gurage Zone to make her HIV-positive status open to the public. Her actions were interpreted as protective of the community, and she was locally identified as the "first female *jegna*." However, she is not the first woman to trespass on traditional gender boundaries. Defiance against existing gendered norms and patriarchal claims is part of the Gurage female tradition, as I describe below.

On the morning of August 2012, Meseret and I watched a pair of male and female plasterers working on the cabin built on a plot of land owned by Meseret's son. The husband stood on top of a ladder, plastering the wall with mud mixed with straw. We were amazed by the wife plasterer's skill of throwing each lump of mud right into her husband's hand. Then, one of the village men in the scene told her that she "was to be punished" on the ground that she "trespassed the male job." From his tone, it was apparent that he wanted to encourage her by implying that she was doing much better than he would expect from a woman. However, the woman instantly responded, "This is the age of democracy. There is no such thing as a male job and a female job." By doing so, she successfully dismissed the man's obvious claim of male supremacy.

The wife plasterer's claim of "the age of democracy" apparently echoed the government's rhetoric for women's empowerment. Compared with its predecessors, the EPRDF, which ruled Ethiopia from 1991 to 2019, was keener to implement this policy. One of its prominent achievements was the Revised Family Code of 2000, which repealed and replaced some provisions of the Civil Code of 1960 that held women in subordinate positions in their marital relations (Teshome, 2002). Moreover, the EPRDF repeatedly demonstrated its commitment to abolishing harmful traditional practices against women, including forced and early marriage (Federal Ministry of Women, Children & Youth Affairs, 2013). However, it is misleading to frame the talk about the woman plasterer's "trespassing" of gender domains entirely as a problem between traditional gender roles and modern women's empowerment. Instead, arguments over gender domains and male supremacy are part of the traditions of Gurage society and elsewhere in Ethiopia.

In rural Gurage society, the division of labor between men and women seems so evident that William Shack, the American anthropologist who published a monograph on Gurage culture in 1966, described that men engaged in "no domestic work," whereas political affairs were "totally men's work." He suggested that domestic work, caregiving, and training young girls for marriage are among women's labor, and assisting in domestic work and aiding the aged and infirm are classified as girls' work (Shack, 1966).

His account of the gender division of labor might seem to leave little room for argument for those familiar with the rural Gurage livelihood, except that the division of labor in some circumstances is less clear than Shack had put it. For example, while Shack classified "watching infants" as young girls' labor, I often spotted during my fieldwork some local boys strolling around carrying their young siblings on their backs.

Moreover, Shack's description prompted a stubborn argument from an unlikely figure. Alemayehu Neri harshly criticized Shack's account of gender division and his whole account of Gurage culture. Alemayehu is a prominent Gurage figure whose father, Kegnazmach Neri, was a local chief under Ethiopia's imperial government. His primary concern seemed to be who should define Gurage culture: a foreign (male) anthropologist or a local (male) elite. One of Shack's statements that Alemayehu found particularly unacceptable was that old Gurage men were engaged in watching infants (Shack 1966). In his book on Gurage culture written in Amharic, Alemayehu asserted that a Gurage elder, "however needy, poor, and without support should he be [...] his task in his old age should not be watching, carrying, or comforting infants" (Alemayehu, 1993, p. 61).

However, this statement seems to describe Alemayehu's firm belief in male supremacy rather than explaining facts about childrearing among the Gurage. If Shack had made any mistake concerning older Gurage men's roles, he failed to perceive the gender norms as open to heated arguments.

YeKake Werdwet

In his claim for the authenticity of the Gurage patriarchy, Alemayehu confirmed the arbitrary nature of gendered norms in his society—without intending to do so. To what extent could gendered norms be challenged by women within the context of traditional Gurage society? This question evokes the image of a legendary Gurage woman known as YeKake Werdwet, who is believed to have lived in the nineteenth century.[10] According to oral tradition, she was an extraordinarily beautiful woman, born to one of the most prominent Gurage families.

> When Werdwet came of age, her father gave her to Abegaz Ferchiye, a man of great fame and wealth. He was a prominent *jegna* at the time. Although the woman was initially happy with her marriage, her happiness gradually eroded because the man married many women. One day, Werdwet stood in front of the elders' assembly, held at their meeting place under the shade of a tree. The elders were startled because their tradition did not allow women to do so. To their surprise, she asked an unusual question: If our custom enables a man to choose his wife and to have many wives, why can a woman not do the same? As the elders were not ready to answer such questions, they gave her an appointment for another day. On the date of the meeting, Werdwet appeared with many fellow women. This made the elders fear that, if this went on, women would turn to public affairs and abandon their household business. To avoid further confrontation, they gave Werdwet an appointment for the third time. Meanwhile, the elders ordered the village men to pull out some enset plants in their backyard on the third appointment day. Their conspiracy was successful. On finding the plants on the ground in their backyard, the women felt obliged to start processing them before they spoiled. As a result, no woman could appear at the third appointment except for Werdwet herself.

Once matured enset plants are cut down, the tiresome labor of getting them ready for consumption is reserved exclusively for women.[11] Although enset is a close relative of the banana, its fruit is unsuitable for consumption. It should first be divided into parts consisting of the bulbs, stems, and leaves. Then, a woman places a stem, usually taller than herself, on a wooden board leaning against a standing plant of enset and shaves it with a sickle. One of their legs is always placed on the stem to fix firmly while shaving. The bulb should also be stripped off its outer part until its edible core is revealed before it is cut and boiled for consumption. Meanwhile, the starch extracted from the stems, which is more considerable in amount, takes much more time and effort because it requires a long process including fermentation and removing fibers before being baked and served as food. Each process requires significant time and manual labor.

The apparent message conveyed by Werdwot's story is that women cannot challenge men because the latter surpasses the former in wisdom. However, Werdwet was not beaten by the elder's conspiracy. The story continues:

> With no fellow women standing with her, Werdwet repeated her request that women should be able to marry many men. This time, the elders turned to her husband, Abegaz Ferchiye, and urged him to concede to her. They told him that the only way to stop the woman was to treat her as exceptional: they had to allow her to marry many men while prohibiting other women from doing the same. Abegaz resisted but finally accepted the proposal, convinced it was the only way to defend the social order. Werdwet, too, accepted the decision and, with her servants, started to tour the households of Gurage men known to be war heroes. She stayed at a household for several days, weeks, or as long as she wanted before heading for another brave man. However, she was able to do so until she was struck down by lightning and died.

Lightning manifests divine power in the traditional Gurage religion (Shack, 1966, p. 109). Werdwet was finally punished with death because she challenged the legitimacy of the rule of male elders. While the story's message is supposed to confirm male supremacy, it also carries contrasting messages. The elders were afraid of women's ability to threaten order. Such fear led to an odd consequence: Werdwet's husband was obliged to concede to her bewildering request despite being the most prominent man in their society.

Moreover, the story has attracted a number of reinterpretations. Bahru Zewde, an eminent Ethiopian historian of Gurage origin, mentioned

Werdwet as a woman who struggled against the male-dominant norms of traditional Gurage society (Bahru, 2002). Another example is an article that appeared in a regional magazine in 2005. It avoids mentioning divine punishment and interprets the story as a critique of discrimination against women.[12] Such reinterpretations, I suppose, should not be framed within the "traditional versus modern" dichotomy. Comprehending these rereadings of Weldewot's story as part of modern emancipatory interventions against the traditional gender norms risks underestimating female agency within the context of Gurage society. Instead, the tellings and retellings of Weldewot's story signify the continuous negotiation over gender norms within the Gurage culture, which have never been short of imagination about alternative forms of sociality.

The legend of Werdwet is about a woman who decisively trespassed the gender boundary and was eventually punished or praised for doing so. Moreover, it is a story about disrupted sisterhood under patriarchal pressure. One may comprehend the story as an eccentric legend that ultimately reinforces the rule of the elders. I disagree. I believe that Ethiopia has an abundance of "culture of defiance" and aspirations for alternative moral engagement. I will demonstrate some of their manifestations in the next chapter.

Notes

1. Anonymous article, 2007 (1999EC) "Mignote bahil wegun akibire ajyet mebbal nebber" *Ilifign* No. 4, p. 49.
2. The Ethiopian calendar is seven years behind the Gregorian calendar. The country entered the third millennium on September 11, 2007, which was the first day of the year 2000 in the Ethiopian calendar. The gap results from an alternate calculation in determining the date of the Annunciation.
3. Data obtained from the Gurage Zonal Health Bureau.
4. UNAIDS estimates that the number of new HIV infections in Ethiopia peaked in 1992 and has been declining since.
5. The GPSDO was first established in 1962 as the Gurage Road Construction Organization. Since then, it has helped construct more than 450 km of all-weather gravel roads, in addition to dozens of elementary

and secondary schools, while also consolidating support from communities and national and international development agencies (Nishi, 2008).

6. The ratio of unmarried people among all testees was not provided. A considerable portion is assumed to be unmarried, as a rigorous premarital campaign was underway during those days.

7. Based on the audience comments on my presentation at the 18th International Conference of Ethiopian Studies (Nishi, 2012).

8. Here, I use the term "unintended effect" after James Ferguson (1990), who discussed the unintended outcomes and effects of development intervention in Lesotho, drawing on the Foucauldian theory of power.

9. It is not my intention to suggest that local male elite-led social movements necessarily have such "anti-politics" effects. Since its establishment, the GPSDO has been active in promoting economic redistribution among the urban and rural populations of the Gurage (Nishi, 2008).

10. Based on my interview with Ato Awlachew Shumneka on August 18, 2010. Sources do not agree about the time she lived. The following story of Werdwet's life and death is also based on information from Ato Awlachew.

11. I once attempted to learn how to process an enset bulb in the backyard of a Gurage household. A neighboring woman spotted the scene and burst into laughter. However, she would have been furious had she seen her son (instead of a foreign researcher) engaged in the same task.

12. Anonymous article, 2006 (1998EC) "YeKake Werdwet" *Ilifign* No. 3, pp. 41–44.

References

Alemayehu, N. W. (1993). *Aset: Yebahil na yetarik meseret (Enset: The foundation of culture and history)* [1985EC]. Bole Printing Enterprise.

Bahru, Z. (2002). Systems of local governance among the Gurage: The Yejoka Qicha and the Gordana Sera. In Z. Bahru & S. Pausewang (Eds.), *Ethiopia: The challenge of democracy from below* (pp. 17–28). Nordiska Afrikainstitutet; Forum for Social Studies.

Biehl, J., Good, B., & Kleinman, A. (2007). Introduction: Rethinking subjectivity. In J. Biehl, A. Kleinman, & B. Good (Eds.), *Subjectivity: Ethnographic investigations* (pp. 1–23). University of California Press.

Borrell, J. S., et al. (2020). Enset-based agricultural systems in Ethiopia: A systematic review of production trends, agronomy, processing and the wider food security applications of a neglected banana relative. *Plants, People, Planet, 2*(3), 212–228.

Federal Democratic Republic of Ethiopia. (2000). Revised Family Code Proclamation No. 213/2000, Federal Negarit Gazetta Extra Ordinary Issue 6/1, 1–9?.

Federal Ministry of Women, Children and Youth Affairs. (2013). *National strategy and action plan on Harmful Traditional Practices (HTPs) against women and children in Ethiopia*. Federal Ministry of Women, Children and Youth Affairs, Federal Democratic Republic of Ethiopia.

Ferguson, J. (1990). *The anti-politics machine: Development, depoliticization, and bureaucratic power in Lesotho*. Cambridge University Press.

Nishi, M. (2008). Community-based rural development and the politics of redistribution: The experience of the Gurage road construction organization in Ethiopia. *Nilo-Ethiopian Studies, 12*, 13–25.

Nishi, M. (2012). Morality, risk, and knowledge in the era of global health: HIV interventions and local responses among the Gurage. In *Paper presented at the 18th International Conference of Ethiopian Studies, Dire Dawa University, Dire Dawa, Ethiopia, October 28–November 3, 2012*.

Sahle, M., Saito, O., & Demissew, S. (2021). Exploring the multiple contributions of enset (Ensete ventricosum) for sustainable management of home garden agroforestry system in Ethiopia. *Current Research in Environmental Sustainability, 3*, 100101.

Shack, W. A., & Marcos, H.-M. (1974). *Gods and heroes: Oral traditions of the Gurage of Ethiopia*. Clarendon Press.

Shack, W. A. (1966). *The Gurage: A people of the ensete culture*. Published for the International African Institute by Oxford University Press.

Teshome, T. (2002). Reflections on the Revised Family Code of 2000 Ethiopia. In A. Bainham (Ed.), *International survey of family law 2002 edition* (pp. 153–170). Family Law.

YeGurage Hizib Ras Ages Limat Dirigit. (2007). *Qicha: YeGurage bahilawi hig, teshashilo yewetta (Revised edition of the Gurage customary law) [2000EC].* Finfine Matemiya na Publishing.
Yeltyopya Kwankwawoch Tinat na Mirmir Maikel. (2001). *Amerigna mezgebe kalat (Amharic dictionary)* [1993EC]. Artistik Matemiya Dirijit.

7

Culture of Defiance

Abstract This chapter turns to the life and engagements of another Ethiopian woman, Asha, whose knowledge of midwifery and the leadership roles she assumed in neighborhood groups were found useful by the local officials and agents who wished to promote community health. They repeatedly sought her services as a traditional birth attendant or HIV home-visit volunteer, neither of which earned her the remuneration she deserved. I argue, however, that the seemingly fragmentary pieces of Asha's engagement in her community point to a sociality centered on care and responsiveness that survived patriarchal impositions and differential systems. Moreover, her tireless engagements with neighbors were underpinned by the "culture of defiance" inherited and shared among Ethiopian women. To shed more light on how Meseret's work is embedded in, but not defined by, local moral perspectives, I present an account of her childhood. Her aspiration to advance through a normative course toward becoming a "caring woman" was complicated by adverse events and troubled relationships. Her recollection of her mother, who died in isolation, offers a focus for her moral inquiry, which emerges at the intersection of crisis, care, and womanhood.

© The Author(s), under exclusive license to Springer Nature
Singapore Pte Ltd. 2023
M. Nishi, *Curing Lives*,
https://doi.org/10.1007/978-981-99-1831-7_7

Keywords Caring relationships · Culture of defiance · Community
health interventions · Patriarchy

In the late 1990s—a decade before I first met Meseret—I visited
Chancho, a rural neighborhood in the Southern Region, with one of my
research assistants. One of our aims was collecting stories about coffee
traders in early twentieth-century Ethiopia (Nishi, 2005). There, I met
Batula, a woman presumably in her 80s at the time. We did not ask her to
share her story because we considered that coffee trading was males' business and did not assume that women had much to say about the topic.
However, on one of the nights we spent in her son's hut and discussed
our research findings, she started to tell a story in the local Silte language,
which I did not understand.[1] One of the men in the scene tried to stop
her. However, my assistant considered her story fascinating and asked her
to continue. On that night, she told us how she made possible a reunion
with her mother, who had left her when she was very young. She could
not tell the year when the reunion happened, but it was probably sometime in the 1930s. The following story is based on the Amharic text,
which my assistant translated from the original story told by Batula.

Batula's mother's name was Kedija. Kedija's husband died shortly after
the birth of their first daughter, Batula. In-laws asked Kedija to marry
one of her deceased husband's brothers, which was a common practice
in rural Ethiopia during those days. However, the brother was much
younger. She protested, saying that she could not wait for years until
he grew up and that she would be an "old woman" by then. She finally
escaped the situation by leaving home for Sidama, a coffee-producing
county in southern Ethiopia.[2] On her trip, she was accompanied by a
male coffee trader. It was the time when Sidama's economy was booming
with emerging coffee production.[3] She settled near Aleta Wendo, a
Sidama town, and became a wealthy moneylender and landowner.

Kedija left Batula behind when she left home, probably because the
daughter was too young to withstand the harsh journey. It was decades
before motor transport was introduced to the southern provinces of
Ethiopia. Kedija had to walk all the way to Sidama, following the caravan
trails. Batula grew up not remembering her mother's face. After she got

married and had children, Batula met a trader who saw her mother during his recent visit to Sidama. "I gave some honey to her [as a sign of courtesy] and talked to her," he said. He then showed a *gabi*—a thick cotton dress to prevent cold weather—and said, "Look, this is the *gabi* that your mother gave me."

I wonder if the man considered Batula's response. She made up her mind that she had to see her mother. During the following days, she became busy preparing for the journey. When everything was ready, she told the trader, "I prepared *shameta*, baked bread, and some cheese at home. Now, let us leave for Sidama." *Shameta* is a thick drink made from barley that boosts one's strength to withstand harsh conditions, such as an arduous journey. However, the man refused, claiming people would arrest him as a kidnapper if he accompanied a woman. When he set off for Sidama, he ensured that Batula did not know when he was leaving.

"I cried and cried," recalled Batula. This situation compelled Batula's husband to propose taking her to Sidama. They set off on the 200-km journey on foot, tracing the path traveled by Batula's mother, with their equipment packed on the back of a mule. However, when they reached Alaba town, 70 km south, she felt exhausted. However, she regained courage when her husband pointed to a church on a far mountain and told her that that was the area her mother lived. "I started to walk like someone who had drunk a large amount of milk," recalled Batula. After that, when she felt tired, she "just looked up the church, and the fatigue went away." However, considering the geographical setting, it was unlikely that she saw her mother's place from where she was. It could be that the husband attempted to encourage her by giving false information. If that was the case, his tactic worked well beyond his imagination.

The couple reached Kase, where they needed to cross seven rivers. Each river was "as large as a *warka* tree," according to Batula.[4] *Warka*, or sycamore fig, common in Ethiopia, is a large tree with a remarkable spread.

When we go into the middle [of the river], the water is high. The water—fwa, fwa, fwa.[5] When we cross the water, there is another. Our mule knows the way very well. [It crosses the rivers] with its nose wide open, snorting, fwa, fwa, fwa.

After crossing these rivers, Batula found her husband fatigued. So she was the one who pulled the mule and led the journey. Some male travelers saw her and said, "This woman is as vigorous as a man. How well she is!" The couple continued even after the sun was down, helped by the light of the full moon—a surprising remark given the lack of security in the provinces in the early twentieth century. They would have heard stories of robberies and killings awaiting helpless travelers. It could be that they joined a large caravan hurrying to Sidama. After a couple of days, they reached Kedija's village and told the locals who they were. Hearing that, they started screaming, "Here comes Batula, here comes Kedija's child!" Villagers began preparing for a feast—they baked bread, filled large pots with milk, and slaughtered sheep. "Eating and drinking, we spent a month with her before we came back," said Batula.

Batula's recollection represents some of the women's strife against their fate against the backdrop of the socioeconomic settings of early twentieth-century Ethiopia. Her journey to Sidama reaffirmed the "culture of defiance" shared along the maternal line. What impressed me that night was how Batula's daughter contributed to performing the recollection. When Batula paused, the daughter continued the story. At times, their voices and gestures were synchronized. The daughter would have heard the mother telling the same story repeatedly to the extent that she could present the scenes as if it was her story. I was listening to a part of the affective memories shared along maternal lines and played out over generations.

In the years to come, at least two other Ethiopian women shared no less compelling stories with me. Meseret is one of them. The other is the late Asha Bule, an *awalaj* or midwife in Addis Ababa. In this chapter, I present Asha's story about her life and her mother before returning to Meseret's story.

Like Meseret, Asha was profoundly concerned about locally shared values of care. However, her stories do not conform to the cultural image of a "caring woman." Instead, her stories point to Ethiopian women's continuous struggles to enact the socialities they envisioned. The stories of Asha and Meseret allow me to explore further the actions of Ethiopian women to survive the patriarchal impositions and differential systems and cure the lives of others and their own. By doing so, I intend to relate their stories to what Didier Fassin (2009) referred to as "another politics of life," which engages the experience of individual human beings and gives place to cultural interpretations and moral decisions.

Moreover, it is my hope to pursue this aim without undermining the plurality of women's voices in Ethiopian society. In her reading of Buchi Emecheta, the Nigerian writer whose works concerned the problems of womanhood, Donna Haraway had it that "women's experience [did] not pre-exist as a kind of prior resource, ready simply to be appropriated into one or another description" (2013, p. 113). She further argued that to think of women's experience was to engage in a "politics of experience" that searched for specificity, heterogeneity, and connection through struggle (p. 109).

A Traditional Birth Attendant

Tekle Haymanot Church, named after a thirteenth-century Ethiopian saint, is one of the eminent Orthodox churches that serve as prime landmarks in Addis Ababa. It stands on the edge of higher ground adjacent to Mercato, the city's largest market center. To the church's south is a valley carved by a stream that flows from the hill. The valley accommodates one of the known poor districts in the city. Most residential structures are *kebele bet* or municipal housing. They are poorly maintained; some of them are barely standing. Still, these housing units have provided shelter, which is the scarcest asset in the vast city of Addis Ababa, particularly for its poor population.

This area has been subject to large-scale interventions conducted by international and local NGOs. Mother–child health was one of the many components of the intervention. In the 1990s, employing traditional birth attendants (TBAs) to supplement modern medical infrastructure, which barely existed in the area, was considered an essential part of such an intervention. A TBA, in this context, means a midwife without modern professional training. In Ethiopia and elsewhere in the Global South, they were involved in community health activities after completing a basic training course (World Health Organization, 1992).

Among those TBAs was Asha Bule, who was in her late 50s when I first met her in 2000. She lived in one of the municipal housing units with her two sons and a daughter. She shared insightful stories on the state of life in Addis Ababa until she died in March 2016 from pancreatic cancer. Her recollection included her experience as a midwife, community health volunteer, and local burial association member. She was daughter to a woman who lived an extraordinary life comparable to YeKake Werdwet's (see the previous chapter). When Asha spoke, her voice powerfully conveyed her pride and conviction as a woman who dedicated much of her life to "curing the lives of others." However, her stories were almost always overshadowed by bitterness, reminiscent of her life at Addis Ababa's margins.

Asha learned midwifery skills from her mother. She had helped women neighbors give birth since the 1970s. Like many other midwives in her country, she did not ask for money for her service. However, her clients were expected to reward her indirectly, often in kind. In the 1990s, Asha was one of the 16 TBAs in the neighborhood who received a month-long training course and were subsequently certified to practice. The course taught Asha and her colleagues about the roles of TBAs and, most importantly, their limits. For instance, they were instructed not to manage complicated cases but to refer them swiftly to the local clinic. They were also given basic instruments for safe delivery.

However, according to Asha, "they changed their minds" after several years of operation. The team of 16 TBAs was dissolved. I wonder if someone told her why they changed their minds—or who "they" were in the first place. While she took her role as a TBA very seriously, she was barely informed of the ways of producing and enforcing knowledge in international maternal health.

Whether TBAs have a role in reducing maternal and infant mortality was an issue of debate among international reproductive health experts. However, by the end of the 1990s, the lack of clear evidence led them to become very pessimistic about the impact of TBAs on promoting reproductive health (Bergström & Goodburn, 2001; Goodburn et al., 2000). In the early twenty-first century, the WHO wanted to consolidate its policy to meet the Millennium Development Goals (MDGs), among which was to achieve universal access to reproductive health and significantly reduce maternal mortality by 2015. In 2002, a WHO document titled "Global Action for Skilled Attendants for Pregnant Women" stated that the organization would promote "skilled" birth attendants or those who underwent professional training (World Health Organization 2002). Thus, TBAs were effectively excluded from the health workforce mobilized for achieving MDGs.

Visiting House to House

Sometime after the dissolution of the TBA team, Asha and her ex-colleagues were approached by the same NGO—this time to be trained as door-to-door HIV education promoters. Some of the 16 TBAs had either died or left the area. However, several women, including Asha, agreed to take responsibility. "Let me tell you why they called on us at that time," said Asha. "If young people had gone to deliver the education, people would have refused them." Before the rollout of free ART in Ethiopia, no one in her vicinity was ready to talk about AIDS, which was believed to be a cursed illness. However, Asha and her colleagues expected people to respect their seniority and previous contributions to the community. Moreover, they were confident they knew how to talk to people. Asha explained how her team communicated with a woman with HIV in the community.

> During our house-to-house visits, [the woman] responded with such words as "What are you?" or "What do you want?" We replied, "Please do not talk to us that way. When someone comes to your home, it is our culture to say, 'our home is for guests' and 'please, come in.' Why are you so angry? What did we do? We came to converse. Let us enter and converse with you. Now, won't you allow us to?" Then, she let us in, but she was embarrassed. She asked, "Is it only my household [you are visiting]?" as if she was talking to policemen who came to search her room. Then we told her, "No, watch us when we go out from here. We will visit all the households. We are here because they sent us to convey information, not for any other business. Don't be offended." And it was after several visits that she confided in us and told us that she had the test

and learned her status [as being HIV positive], to which we responded, "Don't be afraid. You don't need to be ashamed."

"During those days, those who tested positive did not want to talk to others, did not want others to talk to them," continued Asha, "I say, we were so serious at that time. We did not leave any household unvisited. We knocked on the doors one by one, so they would allow us in."

The number of patients was increasing, and many of them became bedridden. Asha remembered that a woman from Chicago guided her team on how to care for terminally ill AIDS patients. The team now became HIV home-visiting volunteers. They started to visit households to tell family members how to change patients' bed sheets and wash their bodies and hair. "I used to tell them, 'Do not stay away. Put gloves on. Do it like this, like this, like this,'" explained Asha. In those days, few people were willing to touch a patient with advanced AIDS. "When a patient died, no one wanted to get close to the body. We were the ones who were called upon to wash and enshroud it." One day, they visited a patient to find that she had just died.

Everyone stayed away. Her sister stayed away. So we undressed her, told them to bring utensils, and told them to get a bench. We placed her body on it and rinsed her thoroughly. Then, we spread the shroud on her bed and wrapped her. It was when we finished everything that the sister came closer, crying, "My mother's child! My mother's child!"

The scene was frustrating because if the woman called her that way, she should not have let others wash her body.

However, "things changed, all those things changed," continued Asha. "The medicine came. They started to take the medicine. Things got better. Things changed." After the rollout of free ART, "patients were not hiding anymore," said Asha. This was when "they" decided that younger volunteers should replace Asha and other ex-TBAs. Asha was so embarrassed that she complained to the project officer if they were to "drop us and look for others." However, according to Asha, the answer was, "What can we do? We were asked to recruit educated ones. You are not educated." One of Asha's colleagues appealed to the municipal office, assuming that they were the ones who requested to recruit "educated ones." However, again, the answer she got was "*igna ayiddellenim*," meaning, "it is not us."

Asha continued to be involved in HIV interventions in her area until she finally stopped in 2008. One of those interventions was the promotion of antenatal checkups at local clinics. Despite the government policy to provide antenatal care for all pregnant women, many women refrained from visiting clinics "in order not to be asked to get tested," explained Asha. She was expected to convince pregnant women in her vicinity to visit nearby clinics for both antenatal care and HIV screening.

> "There was a pregnant woman I knew," Asha recalled as she started to tell me how she convinced one of her clients. The woman had given birth to five children, but she had never gotten tested because she had always given birth at home. One day, she found Asha on the street and said, "Akkoye (Asha's nickname), please attend to me when I give birth." Asha answered, "I will attend to you only if you go to the clinic, get tested, and show me the certificate." The woman was surprised and asked, "Do I have to get tested?" Then, Asha said, "What is the matter? Why are you afraid? Get tested, and if you are positive, you will be treated and give birth to a healthy baby. If you are negative, they will give you instructions so you will not get infected in the future."

The woman went to the clinic and told them that Asha had convinced her. "They were so happy because I convinced the woman who had never consulted the clinic," said Asha. "I was awarded for sending many women [to the clinic]. They gave me an umbrella and a bag." Furthermore, "My name went all the way to America," said Asha, by which she meant that her name was mentioned in the project report they sent to donors in the US. She then added, "*Gin ine iziw negn*," meaning that she remained where she was. "I did all this, but what did it mean? Now, I am in the middle of a wasteland because I am not educated. Because I am not educated, they threw me away," said Asha, "But I do not regret much. It was for those around me." She added, "*Bizu hiwot adigneallehu*," meaning, "I cured many lives."

The Amharic verb *adane*, which I translate here as "to cure," shares the same root as *medhanit*, meaning medicine or remedy. The verb may also be translated as "to save." To cure or save someone's life is a familiar expression for Amharic speakers. One may save someone's biological life by treating them when critically ill. One may save another's social life by offering relevant *mikir* or guidance during a moral crisis. Asha's claim of curing lives applies to both contexts. As a midwife and TBA, she saved the lives of mothers and newborns who might not have survived difficult deliveries. As an HIV home-visiting volunteer, she was engaged with patients whose dignity and sociality had been deeply wounded.

I am not suggesting that the community-based interventions in Asha's neighborhood were irrelevant. They were there to serve the marginal population. Their methods were shifting because the public health targets were shifting. It is no surprise that they found someone like Asha useful to pursue their targets with their limited budget. Nonetheless, I wonder how many of them were aware that they eventually left her "in the middle of a wasteland" with a bunch of certificates of appreciation in her hands. They found her useful when there seemed little hope of curing—biologically or socially—the lives of those dying of AIDS. They replaced her with "educated ones" when antiretroviral drugs made the situation easier to handle.

Asha's story obliges me to ask if her "curing lives" was tradable with the antiretrovirals, which Ethiopians refer to as *yedme marazemiya medhanit* or "life-extending" medicine. Her aim was not to extend her clients' biological life span but to engage with the lives and deaths affected by AIDS. She cleaned and wrapped a deceased client's body because she believed being treated like anyone else after death was an indispensable part of moral life.

Asha's Mother

Asha's mother came to Addis Ababa during the Italian occupation (1936–1941), and in the subsequent years, she married several times and adopted Asha and her younger brother. "She did not give birth to us. She was barren," explained Asha. She was very young when she was adopted. "It is her whom I know [as my mother]. It is her whom I call 'my mother.' They used to call her by my name. They called her Adasha, meaning Asha's mother."

Asha's mother was an Oromo woman born to a Christian family in a village near Sendafa, northwest of Addis Ababa. Her name was Desu. Her father, Gamba, was among the most prominent elders in the vicinity. He was very fond of Desu, who was raised and wed her Christian husband there. However, after bitter quarrels with her husband, she ran away from home. (Note that what is explained as "bitter quarrels" here might mean physical violence from her husband.) She came to Addis Ababa, where she married a Yemeni merchant, converted to Islam, and changed her name to Fatuma. She did so because she learned that her first husband was after her and decided that he would take her back home by any means. She expected her Christian husband would stop seeking her once she converted to Islam. That was why she asked a Muslim Oromo elder in Addis Ababa to witness her conversion and arrange a marriage with an Arab person.

However, she soon found it difficult to communicate with her Yemeni husband because he did not understand her mother tongue, Oromiffa (the Oromo language). He spoke fluent Amharic, the lingua franca in urban Ethiopia, but she did not. She had to leave him. She married another Yemeni, who took her to Mojo, a town some 70 km to Addis Ababa's southeast. The town was known for its high prevalence of malaria, which soon afflicted Desu. On hearing her critical condition, her father came on horseback to take her back to his village. The local clergyman who christened Desu visited their household and was startled to find that she had converted to Islam. He complained bitterly to her father. When Desu got better, her father tried to convert her back to Christianity. However, he did not insist after she refused his request. "Her father was such a tender-hearted person," explained Asha.

Desu returned to Addis Ababa and started to live with a Yemeni spice trader, Sala. Desu adopted Asha when she was with him. "I do not remember how old I was, but I could be four. I still see fragments of dream-like memories of my life in his house, even though I was very young," said Asha. There was an Oromo family in her neighborhood, and the wife used to invite Desu for coffee. There she attracted the attention of Grazmach Bule, an Oromo nobleman who frequented the same household. He was from Arsi, a province to Addis Ababa's south-east predominantly inhabited by Muslim Oromos. Sala, Desu's Yemeni husband, said to her, "I am only a spice merchant, but you are tall and glamorous. You are made for a rich man. Now I shall let you go." Asha said that she did not know why he said this to her. "It might be that he was suspicious [of the relationship between Desu and Bule]."

Asha remembered the day she moved into Bule's house. "I annoyed my mother because I kept saying, 'Let me go back to Sala's place. Sala is my father.' Later, one day, she said to me, 'Sala is not your father. I lied that he was your father. Your real father is this one,' to which I conceded. Children were naïve during those days." Asha remembered that Bule, her new father, was very fond of her. They lived a luxurious life. "I had seven spare dresses in the wardrobe when someone who had one spare was considered rich." Meanwhile, Desu had to leave the house when she adopted a boy against her husband's will. He was unhappy that the boy was from her kinspeople in north Shewa. Elders were summoned. They told her that if she wanted to raise a boy, she had to choose one from among her husband's kin. She refused. Humiliated, the elders turned to the nobleman and told him to choose between the woman and his kinspeople. "Give me time," he begged them. "I will make sure that she will turn the boy away." Then, he pleaded with her, "Give up the boy. Let us not break our marriage," to which she answered, "No way." He divorced her.

When Desu left the household, she hired four mules and a pack-horse to transport her belongings to her home village. However, all those precious possessions—cash, golden bracelets, silver ornaments of all kinds, and the dresses, including Asha's seven spare dresses—were stolen by men who visited Desu's household in the guise of guests. They identified themselves as the nobleman's relatives and said they were there to mediate reconciliation. Sheep had been slaughtered to offer them a feast. They ate and drank and waited until everyone was fast asleep. Then they left with all her belongings before dawn.

Desu and her children (Asha and her younger brother) moved again to Addis Ababa and started a frugal life, dependent on a cow she brought from the village. They lived in Sidamo Tera, a crowded quarter in Mercato. However, neighbors complained that her cow attracted many flies to the extent that "they could not sleep." The family lived there for years but finally moved to a tenement house behind the Tekle Haymanot Church. An Arab merchant owned the property, which was very close to the neighborhood Asha later lived in, and served the community as a midwife, as described in the previous section. There they had good access to an open meadow to graze the cow. However, Asha remembered that Desu was engaged in frequent fights with neighbors during those days.

Asha gave birth to her first son when she was 16 years old. The father of the son was a wealthy Ethiopian trader. Then she left home to live with another man, with whom she had several children. However, she later returned to live with her mother after a bitter quarrel with her husband. Like her mother, Asha did not stick to a marital union. Her mother, Desu, died in the tenement in 1973, four years after Desu's father, Gamba, died in his village.

After Desu's death, Asha continued to live in the tenement house with her children. In 1974, amid civil protests against the government's failure to respond to the great famine in the northern provinces, the military committee took over power and executed Haile Selassie I, the last emperor of Ethiopia. The following year, the Provisional Military Administrative Council (1975) proclaimed the nationalization of urban properties. The tenement house Asha lived in was confiscated and turned into municipal housing. The occupants, including Asha, continued to live there. Several years later, the government decided to build a school in that area. Asha and her neighbors were relocated to temporary shelters built on a swampy meadow where her mother had grazed the cow. Asha explained how the neighborhood struggled with the muddy water that inundated the area during rainy seasons. They collected stones and threw them into the muddy soil for years until the land became more or less negotiable. The permanent housing that government officials promised never materialized because, according to Asha, they "ate" the money. Asha eventually raised three of her children in a temporary structure.

Life at the Margins

Asha spent much of her life in the municipal housing located at the southern edges of Mercato, the vast trading center of Addis Ababa. The streets of Mercato are packed with warehouses, wholesale stores, retail stores, merchant inns, cafes, and restaurants that serve meals prepared for Muslims or Christians. Although the area has undergone massive redevelopment in the past decade, some parts have retained their shape from the mid-twentieth century. Particularly fascinating is the area adjacent to Merab Hotel, the five-story building that has served as one of the market's landmarks. Along the street descending the slope southwards, retailers sell modern and traditional household utensils, building tools and equipment, local horse gear, and other craft items. Narrow lanes branching from the street lead to stalls selling spices, grains, and fermented enset starches from the south. Further down the street were numerous workshops where used cans from around the town are repurposed into kerosene lamps or pans for roasting coffee beans. Near the end of the street is Chid Tera, or the "Straw Corner." Suburban farmers load their donkeys with piles of straws and bring them to this corner. Those straws are mixed with mud and used as building material. The street ends at the point where it meets the avenue that marks the southern boundary of Mercato. This part of the avenue is known as Berbere Berenda, meaning "pepper terrace," because both sides of the avenue are occupied by wholesalers of dried *berbere* (a kind of red pepper), an essential spice for local daily meals.

Berbere Berenda is located along the top of the terrace that stretches from the Tekle Haymanot Church toward the southwest. The municipal housing once occupied by Asha and her children is found in the depressed area behind Berbere Berenda. This area is occupied by those who do not have the capital and network to participate in Mercato's thriving economy. In this area, men make their living by hawking handmade *mewalwiya* (mops) or by being *lewachi* (literary meaning "one who exchanges") that visit households to collect used clothes in exchange for cheap plastic utensils. Women make money by selling homemade *tella* (local beer), *injera* (local bread), or fried dough.

Asha was particularly articulate when I asked her about her experience as a member of *iddir* or burial associations in the neighborhood. In Addis Ababa, where individuals often live among strangers, they participate in *iddir* to avoid solitary death. Such a death is unacceptable because, as an Ethiopian sociology student once noted, the volume of wailing voices escorting the body to the grave demonstrates the love and affection people have for the deceased (Mekuria, 1973, p. 13). "May God not deprive you of people to bury you" is a blessing used by some Ethiopians (Alemayehu, 1968, p. 12). Thousands of *iddir*, each of which has a membership of dozens to several hundred people, operate in Addis Ababa's neighborhoods. Members of an *iddir* are expected to contribute a fixed amount of money every month. When a registered member or one of their close family or household member dies, the association is responsible for burying that person; more specifically, the *iddir* provides the labor, money, and equipment to conduct the burial. By becoming an *iddir* member, even "a daily laborer with no relatives" might claim that they still had someone on their side, as Ethiopian Sociologist Alemayehu Seifu pointed out (1968, p. 14).

Iddir is essential because "no one should die alone," said Asha. If someone dies alone, "the municipality picks up the body and throws it away, like rubbish. That should not happen to any person." Her statement echoes the voice of a deprived woman in 1960s Addis Ababa. Asked why she opted to pay a monthly contribution to her association "when she needed the money very badly," she succinctly answered, "I do not want my corpse to rot on my bed or to be eaten by a hyena" (Alemayehu, 1968, p. 14). Some destitute *iddir* members give up meals to squeeze out the monthly contribution, as was the case for this woman. However, it is wide the mark to say that she was engaged in an overreaching investment in funeral rites despite her deprived status. She was obliged to invest in the association because otherwise, she was deprived of people who would bury her. *Iddir* may be described as a form of insurance to cope with the uncertainty of life, which, in this context, does not mean poverty or premature death. What is at stake here is their death as the final chance to prove that their lives mattered. In Ethiopia, wealthy and well-connected ones with many relatives would not find *iddir* essential because they already have people to mourn their deaths. However, *iddir* is crucial to life at the margins, as was the case for Asha and her neighbors.

Nevertheless, *iddir* activities involve frequent disputes among members, so Ethiopian city dwellers often remark that staying with the *iddir* is crucial evidence of patience toward others. It often takes more than moderate tolerance, as Asha elaborated to me. In 1994, the Laborers' Iddir—this was the name given to one of Asha's *iddir*—was on the verge of collapse when the treasurer misused and lost most of the 15,000 ETB reserved for funeral payment. "Every member who had membership in another *iddir* left this *iddir*," explained Asha. "What is the use of an *iddir* with empty coffers?" The amount of cash left was 300 ETB—only half the amount needed for a member's death. Only those who could not afford to join another *iddir*, which usually required a considerable admission fee, stayed with the impoverished association. They agreed that everyone should contribute 30 ETB, in addition to their monthly contributions, to reestablish their *iddir*. Because Asha did not have the cash, she contributed the amount over the following months.

Among the members of the Laborer's Iddir was an older woman who begged on the street. When she became too ill to move around, she asked a neighbor to deliver her monthly contribution to the association's cashier. However, it was revealed after a year that the man was spending the money for his own sake. An urgent meeting was called. Members were convinced that the woman's membership should not be canceled, despite the strict rule that anyone who failed to contribute for three consecutive months had to be dismissed. They agreed that the only way to keep her membership was to have the money paid. This might sound harsh, but they had learned from their experience that they had to be very strict about money. Unfortunately, the man was so deprived that he could not pay back what he had stolen from the woman. The situation forced the members to agree that everyone should contribute money to undertake her discontinued payment.

Meanwhile, everyone was convinced that the man's misconduct needed to be punished by expulsion until his wife showed up to appeal for reconsideration. Her problem was that she would lose her membership status if the *iddir* expelled her husband. She explained that although they were registered in her husband's name, it was she, and not her husband, who was paying for the association contributions. She insisted that it was unfair that she should be punished because of her husband's misconduct. Another meeting was called, and it was proposed that the wife "inherit" her husband's membership (as if he was a "dead" member). Some members resisted this idea on the grounds that if she inherited the membership, they would eventually be responsible for burying the mischievous man because the association's rule would force them to bury members' spouses. However, according to Asha, they agreed that the wife should inherit membership. The decision was right, said Asha, because they knew the wife would have little chance to join another association. They knew that she could not afford the admission fee, and if she could, her admission might still be refused, because they rarely admitted someone who had been expelled from another *iddir*—and she would die alone.

Asha's consistent disagreement with unattended deaths best signifies the nature of her moral engagement with others. For Asha and her neighbors, death marked the height of the relationships they managed to maintain amid social and economic distress. Asha insisted that *iddir* was essential to life at the margins because it called up mourners to attest that the deceased's life mattered.

The life trajectories of Asha and her mother, Desu, attest to the women's endeavors for "curing lives" or resisting systematic alienation of values and relationships that shape their lives. Desu's struggle to make sense of her own life was so often disrupted amid patriarchal social impositions that she eventually found herself and her daughter Asha at the margins of Addis Ababa's socioeconomic landscape. When Asha became a grown-up woman, her engagement with her neighbors was found useful in promoting community health schemes with shifting foci. Her contribution to those schemes was underpinned by her continuous engagement with the neighboring people as a midwife and devoted member of burial associations. Those health schemes exploited Asha's long-term investment in her community. However, as a community health volunteer, she was always regarded as disposable and exchangeable. She was never given a chance to have her say in the production and application of global health knowledge. Asha's inclination for benevolent actions might seem to contrast with her mother Desu's apparent focus on the state of her own life. However, I suppose that the stories of the two women constitute two sides of the same coin, representing a "culture of defiance" shared along the maternal line.

Meseret's Early Life

In August 2009, Meseret gave birth to a daughter. Three women from her home village visited to celebrate the occasion. They presented a pair of clay pots, one filled with cheese and the other with butter. Two of the women represented her relatives. The third woman was not her kin, but she wished to use this opportunity to respond to what Meseret had done for her previously. One of this woman's sons spent several months in prison at the edge of Welkite town due to a dispute among relatives. Because the mother could not visit the son often enough, Meseret visited and provided him with food and water, which helped him cope with the facility's harsh circumstances. That was why the woman joined the occasion to provide Meseret with butter and cheese, which are considered essential to recovery after birth.

When I visited Meseret in the same month, she was lying on her bed. With her daughter in her arms, she started talking about her childhood memories. She lived in the village we visited the previous year (see Chapter 6). One day, she had borrowed a thick baggy coat and a big hat owned by a man hired to cut the grass in her family's backyard. She went to the meadow and put on the oversized coat and hat. She then climbed a tree and, from there, jumped on a horse owned by one of her neighbors. She trotted the horse through the meadow until she thought she had gone far enough. She abandoned her horse and returned home. Meanwhile, the owner noticed that his horse was gone. Villagers took part in the search for the lost horse. To their horror, a man reported seeing the horse trotting through the meadow with a *seytan* (demon) on its back. "I was such a mischievous girl," she said. "The memories of my childhood always amaze me."

This story made a strong impression on me not only because of the great mischief she did to the villagers—I did not ask them what they thought of it or if their memory matched hers—but the character of the young Meseret was very different from what I learned from other parts of her childhood story. Most of the childhood memories she shared with me pointed to a girl with a "reserved" personality, as shown below. Of course, she would have acted differently on different occasions or could have changed her behavior over the years. My question is, why did this particular story come to her mind at that specific moment? Perhaps, she wanted to tell the story to her newborn daughter to tell what her mother was like.

Meseret was born in 1984 in a Gurage village. When Meseret was two years old, her father "snatched her" from her mother and took her to his house in Addis Ababa. He was among the Gurage men who owned two homes, one in the home village and another in town, with the latter serving as the base for his business activities. Three years later, Meseret visited the village and saw her mother, who had been divorced by her father and lived alone in her hut. Her poverty was plain to see even for the young Meseret: Her backyard lacked the enset plants for her consumption. When she heard that her father was considering giving some enset to his ex-wife, Meseret quickly asked her father if she could choose the plants. He agreed. Meseret pointed to the largest and finest plants in his backyard until he told her to stop. Those enset plants were designated for special occasions, like hosting his guests. Her mother took them home. Meseret was pleased that she did something useful for her mother for the first time. However, she felt sorry that she was a girl who "could not be as much help as a boy" to her mother, recalled Meseret.

Some years later, when she visited the village, Meseret learned that her mother had died. "I hated my father," said Meseret, for he had been trying to hide her mother's death. It was unthinkable for Meseret that a mother would be buried without her daughter in attendance. She decided to stay in the village when her father returned to Addis Ababa.

She had not stayed long in the village when one of her uncles died. On the evening his body was buried, Meseret was among several girls who were busy serving coffee for some funeral attendees who remained in the deceased's household after the others had gone home. She noticed that they started to look at her. She heard one of them say, "Whose young

cow is she?" Then, people began to whisper. Meseret did not understand what was going on. Because she grew up in town, she was not well
versed in the manners and procedures in the village. Later, she learned
that those people were the parents and relatives of the man to whom
she was betrothed. They wanted to see her before the formal proposal of
marriage.

After her father returned from Addis Ababa, the elders representing
her future husband's lineage came to ask for his decision. As her father
had been away for months and was unaware of this development, he
was surprised and asked them to reconsider, saying, "Please, she is only
a child." However, they insisted, saying they would foster her until
she became a grown-up woman. This statement convinced her father.
After several months, a relative of her would-be husband brought some
clothing for her. On seeing them, the women in her house were upset
that those clothes were "totally out of fashion." However, Meseret did
not look at the dresses or say a word. She believed a would-be bride had
to keep silent about the marriage procedures. Otherwise, she would have
developed a bad reputation, and village girls would have sung songs to
tease her for her misconduct, explained Meseret.

Meseret thought she had to earn money to prepare for her wedding.
She wanted to be well-prepared for her marriage. She began to bake
injera and sell them at a local market. However, her stepmother repeatedly discouraged her from doing so. Perhaps the woman did not want
to lose her business, as she was also selling *injera*. Despite the tempta-
tion to talk back, Meseret again chose to remain silent. However, when
her relationship with her stepmother deteriorated, she decided to go to
Addis Ababa to live with her father. Once there, she visited the household
of one of her maternal uncles to find comfort. Her cousins were kind
enough to buy the things she needed for her wedding. A week before her
marriage, she left her home in Addis Ababa. As her father had left the
village several days earlier, she had to travel alone. When she reached the
bus terminal, she was robbed of all her belongings, including the ornaments for display at her wedding. She was so despairing that she decided
not to return to the village.

Those waiting for Meseret at the village were very upset when she
failed to appear for her wedding. The bridegroom's family had already

arrived. Humiliated, they refused to eat anything provided by the bride's family. They even refused to drink water. Her father rushed back to Addis Ababa to find her at home. It was only after he tearfully begged her to do so that she agreed to go to the village with him. The villagers were so delighted that they kissed her and sang for her: "O bride, are you back here? Are things fine with you? You stayed in Addis Ababa, leaving your hero alone." Again, she stayed silent. She did not explain anything because she thought she ought not to.

The following year, Meseret gave birth to a baby girl who died after two months. Her death did not cause significant emotional pain for Meseret, who explained, "Maybe it was because I was immature at the time. I did not feel any pain because I could not foresee any good thing then. I was too worried about raising her."

Some of her in-laws were mean to her from the beginning, according to Meseret. Later, they became increasingly unhappy with her for reasons she could not understand. Her husband was initially very fond of her but later changed his attitude. One day, they told her to send for her family. Her father and several male elders of her lineage came to discuss the matter. Although she believed that she and her husband loved each other, she told them that she could not continue to live with them, that she wanted to go home, and that she did not want anything from them. As soon as she returned home, her family received many marriage offers. This bothered her, and she left her village again to live in Addis Ababa.

While living in town, she met a man whose home village was close to hers. She accepted his marriage proposal. Although she hoped to settle in Addis Ababa, he took her to his village. After spending some time with her, he returned to town. Meseret was left alone. Her husband's parents had already died, and her new relatives-in-law were not kind to her. They were unwilling to help her when she prematurely gave birth to a boy.

When her son started walking, her husband was called up for the border conflict with Eritrea. Upon hearing this news, Meseret rushed to Addis Ababa to see him. He calmed her down by telling her that everything would be fine, that he had arranged for everything, including her ability to receive a part of his salary directly from the government office, and that he would return.

Cure and Defiance

On an afternoon in August 2013, Meseret and I were standing on the meadow in front of the Fana office. A monthly meeting of her association was scheduled on that day. As the association members arrived, Meseret spoke with them individually, asking questions like, "Tadde, how is your Asthma?" "Tsehay, you came here alone? Where is Arganesh? I asked you to bring her along with you." Arganesh was a woman who had severe auditory difficulty. When she arrived, Meseret tried to determine the outcome of her recent visit to a hospital in Addis Ababa. Fana members had contributed money to cover her travel expenses because she had been experiencing some unspecified health problems, which seemed severe. However, Meseret could only learn that many things went wrong at the hospital. Since Arganesh was illiterate in written and sign languages, Fana members had difficulty communicating with her.

When Meseret spotted a tall man wearing a *taqiyah* (a Muslim skullcap), she asked him whether he remained in the same place he used to live. The town was experiencing a sharp hike in room rents. Some impoverished members, including this particular man, were forced to change places every time the house owner declared a rent increase, usually toward the town's peripheries. Meseret had lost contact with some members who changed locations without informing the Fana office. She did not know whether they stopped showing up because they had left town or because they were too sick to move around.

Meseret started the meeting 30 minutes late, asking the members to share their stories. "It is good to tell your story. By sharing your story, you can reduce your burden," Meseret said. Two women and two men responded to her request, and they spoke in turn. One woman mentioned that her ex-husband had not been responsive to her during some critical moments. The ex-husband was present at the meeting and tried to interrupt, but Meseret stopped him. When he asked if he had the right to speak, Meseret responded, "Let her speak what she has on her mind." The woman finished the story and was met with applause from the attendees.

The scene contrasted Meseret's narrative of her earlier life in which she believed that she was not allowed to explain herself during some critical moments. However, I doubt that a story of sheer transformation from a voiceless young woman to an eloquent leader explains her life trajectory. Instead of comprehending her role as a *jegna* or a war hero in the local fight against AIDS, I suggest that her moral engagement is better understood as guided by the culture of defiance. Though often rendered invisible, it consists of an essential part of female (and male) lives in Ethiopia. Some of the most vivid demonstrations of such defiance I described earlier in this chapter are how Batula and Desu escaped patriarchal relationships to "enact the possibilities they envision" (Rosen, 2003; as cited in Biehl et al., 2007, p. 8). I often observed spontaneous manifestations of a culture of defiance in Meseret's interactions with others. How Meseret guided one of the Fana member women to speak of her frustration toward her ex-husband was an instance of such practice.

In Ethiopia and elsewhere, caring relationships have too often been disrupted within patriarchal social settings. Meseret's recollection of her mother, who died in isolation, offers a focus for her moral inquiry, which emerges at the intersection of crisis, care, and womanhood. Meseret's interactions with some caring women—Murgad Werku (see Chapter 6) and her mother—should have guided her moral engagement with the Fana members. However, I wonder if she could enact the possibilities she envisioned—curing the lives of individuals affected by HIV—without recourse to the culture of defiance inherited and shared among Ethiopian women.

Asha was another woman who claimed to have cured the lives of many individuals. As a midwife and HIV home-visiting volunteer, she engaged with the lives and deaths affected by AIDS. Asha did so as part of her continued engagements with neighbors who lived at the margin of the socioeconomic landscape of Addis Ababa. Her story is reminiscent of home-based care volunteers in the central Mozambican city of Chimoio. They were "motherly" women who were wise and gifted at confronting suffering and death gracefully and compassionately. They were well grounded in the "local logics of care" and were oriented toward the material and spiritual needs of their patients (Kalofonos, 2021, p. 133). However, they felt abandoned and exploited amid the rapid scale-up of HIV intervention.

Asha, too, was a "motherly" woman who dedicated much of her time and efforts to "cure the lives" of her neighbors. However, her actions did not conform to the local norms of traditional womanhood, not because she was actively engaged in "modern" primary health programs, but for her, there was "no moment of innocence" (Haraway, 2013, p. 116) when Ethiopian women lived comfortably within the locally shared sphere of gender norms. The trajectories of Asha and her mother, Desu, attest to women's endeavors to reclaim disrupted life values. Both of them went through arduous journeys. Desu's struggle to make sense of her own life was repeatedly disrupted amid patriarchal social impositions. Asha's engagement with her neighbors was useful but regarded as exchangeable within the community health schemes with shifting foci. Yet, the stories of Asha and Desu attest to the female endeavors played out repeatedly in Ethiopia to enact the values and meanings of life they envisioned.

Notes

1. Silte language is an Ethiopian Semitic language spoken in the Silte Zone, an administrative unit located east of the Gurage Zone.
2. County, or *awrajja,* was an administrative unit that subdivided the provinces (*teklai gizat*) of Ethiopia until 1995.
3. During the early twentieth century, rulers of Sidama and neighboring counties promoted the production of coffee, which emerged as Ethiopia's major export commodity. Charles McClellan (1986, pp. 184–187) described the rise of coffee production during the 1920s in Gedeo, a county adjacent to Sidama.
4. I cannot identify these rivers because this part of the Great Rift Valley lacks any permanent flow of water of the size she mentioned. She probably traveled during the rainy season when the area was covered by flooding water.
5. It is unclear from her narrative whether "fwa, fwa, fwa" represents the sound of water or the mule's snort, as she described in the next sentence.

References

Alemayehu, S. (1968). Eder in Abbis Ababa: A sociological study. *Ethiopia Observer, 12*(1), 8–18.

Bergström, S., & Goodburn, E. (2001). The role of traditional birth attendants in the reduction of maternal mortality. *Studies in Health Service Organization and Policy, 17*, 77–95.

Biehl, J., Good, B., & Kleinman, A. (2007). Introduction: Rethinking subjectivity. In J. Biehl, A. Kleinman, & B. Good (Eds.), *Subjectivity: Ethnographic investigations* (pp. 1–23). University of California Press.

Fassin, D. (2009). Another politics of life is possible. *Theory, Culture and Society, 26*(5), 44–60.

Goodburn, E. A., et al. (2000). Training traditional birth attendants in clean delivery does not prevent postpartum infection. *Health Policy and Planning, 15*(4), 394–399.

Kalofonos, I. (2021). *All I eat is medicine: Going hungry in Mozambique's AIDS economy*. University of California Press.

McClellan, C. W. (1986). Coffee in centre–periphery relations: Gedeo in the early twentieth century. In D. L. Donham & W. James (Eds.), *The Southern marches of imperial Ethiopia: Essays in history and social anthropology* (pp. 175–195). Cambridge University Press.

Mekuria, B. (1973). *Eder: Its roles in development and social change in Ethiopian urban centers* (4th-year senior essay, School of Social Work). Haile Sellassie I University.

Nishi, M. (2005). Making and unmaking of the nation-state and ethnicity in Modern Ethiopia: A study on the history of Silte people. *African Study Monographs,* (Suppl 29), 157–168.

Provisional Military Administration Council. (1975). Government ownership of urban lands and extra houses proclamation no. 47/1975, Negarit Gazeta 34/41, 200–214.

Rosen, L. (2003). *The culture of Islam: Changing aspects of contemporary Muslim life*. University of Chicago Press.

World Health Organization. (1992). *Traditional birth attendants: A joint WHO/UNFPA/UNICEF statement*. World Health Organization.

World Health Organization. (2002). *Global action for skilled attendants for pregnant women*. World Health Organization.

Conclusion

Several years after I concluded my research with the Fana association, Meseret informed me that she had left the office. I expected the news because she had told me several times about her intention to do so. "I wonder why I am concerned about so many people's problems when I could have lived just taking care of myself," she said.

Meseret emphasized that it was her choice to leave the association and live her own life. However, I suppose that the systematic indifference toward her care work with Fana members obliged her to choose between the association and "her own life," though both were equally earnest threads in her trajectory of moral survival. The ART scale-up in Africa gave a chance for survival to millions of individuals, including Meseret. However, it was through her engagements with the Fana members that she reclaimed a moral life that involved "negotiating important relations with others, doing work that means something to us, and living in some particular local place where others are also passionately engaged in these same existential activities" (Kleinman, 2006, p. 2).

This book has addressed the issues concerning pharmaceuticalization by focusing on some crucial moments of the formation and transformation of HIV care in Ethiopian society. It found that the rapid expansion of the ART program in Ethiopia led to a health system in which the suffering of those whose lives were disrupted by HIV was rendered invisible, and the actions to cure the lives of those sufferers were marginalized. The process has been facilitated by global health experts' obsession with the economic value of the intervention and reinforced by the Ethiopian government's excessive preoccupation with controlling the international and domestic resources for health.

However, it also illuminated some local actions that pointed to what Didier Fassin (2009) referred to as "another politics of life," which engages the experience of individual human beings and gives place to cultural interpretations and moral decisions. Meseret and Asha were among the women who cherished the spheres of relationships left unattended by people in power. Both women claimed to have cured some lives through their long-term engagements with individuals affected by the epidemic. Their actions were embedded in culturally shared values and meanings of life and death, as explained in Chapters 4 and 7. Moreover, I argued in this book that their actions were informed by what I refer to as the "culture of defiance" shared by some Ethiopian women. In Ethiopia and elsewhere, caring relationships have too often been disrupted within patriarchal social settings and differential systems. Without recourse to the memories of counterclaims and defiant actions against such impositions, they could not pursue the values and relationships they envisioned.

The caring relationships maintained by those women were found useful in health intervention schemes aimed at bringing about social changes. However, those schemes with shifting foci often exploited those women's capacity and social assets and left them abandoned, as discussed in this book and by Maes (2016) and Kalofonos (2021). That said, it is not my intention to suggest that the "universal treatment" strategy was deemed or intended to promote the economization of life by alienating culturally informed values and meanings of life. When I first met Meseret in 2007, two years after the nationwide free ART program was commenced in Ethiopia, the antiretroviral drugs were powerful agents of change. They cured sick bodies and helped people to overcome their fear of AIDS. They helped Meseret, and other actors in the HIV movement break the silence around the disease and heal the social ruptures caused by the epidemic. In Ethiopia, this process was reinforced by aggressive investment in primary health infrastructure between 2005 and 12 under Dr. Adhanom, Tedros, then the country's health minister.

I discussed in this book how this trend was reversed during the following years. In the city of Welkite, local health institutions referred individuals with multiple health, economic, and social burdens to Fana, the small association of people with HIV led by Meseret. However, this practice turned out to be similar to what Biehl called "triaging care" (Biehl, 2005, p. 22) when defunding non-pharmaceutical HIV care became the new trend. Those whose problems were too complicated to solve with antiretroviral drugs were effectively removed from the scope of care and support that would have helped them to reclaim their lives disrupted by HIV and other illnesses, economic distress, and social exclusion.

Furthermore, the local practice of "triage" was facilitated by health policymakers' excessive focus on the economic value of intervention and away from the individual state of life. Such practices have amounted to a mode of governance that promotes population health without regard to its consequences over the often complicated states of life. Such a mode of governance is reminiscent of what Breckenridge (2014) saw in the administrative architecture of postcolonial African states, and Murphy (2017) referred to as the "economization of life" based on her study on the population health intervention in South Asia. The obsession with the economic value of intervention, a significant driver of pharmaceuticalization, is the tradition in population governance in the Global South that took its shape long before the coming of universal HIV treatment.

However, that is not to say that local actions for health that give place to culturally shared values and meanings of life did not take shape in Ethiopia and elsewhere. It is possible to consolidate broader interventions that promote long-term actions aiming at "curing" the lives disrupted by illness, isolation, and deprivation. Such actions would become robust when guided by local initiatives and inspired by the "culture of defiance" that enables women and men to maintain caring relationships resisting oppression and injustice. Such actions become sustainable when they have access to adequate resources and infrastructure to enact the socialities they envision.

References

ACT UP. (2000) W.H.O. sold out to big pharma. *ACT UP Historical Archive*. https://actupny.org/reports/durban-who.html. Accessed on March 13, 2021.

Abraham, J. (2010). Pharmaceuticalization of society in context: Theoretical, empirical and health dimensions. *Sociology, 44*(4), 603–622.

Adjé, C., et al. (2001). High prevalence of genotypic and phenotypic HIV-1 drug-resistant strains among patients receiving antiretroviral therapy in Abidjan, Cote d'Ivoire. *Journal of Acquired Immune Deficiency Syndromes, 26*(5), 501–506.

Albarracín, D., et al. (2005). A test of major assumptions about behavior change: A comprehensive look at the effects of passive and active HIV-prevention interventions since the beginning of the epidemic. *Psychological Bulletin, 131*(6), 856–897.

Alemayehu, N. W. (1993). *Aset: Yebahil na yetarik meseret (Enset: The foundation of culture and history)* [1985EC]. Bole Printing Enterprise.

Alemayehu, S. (1968). Eder in Abbis Ababa: A sociological study. *Ethiopia Observer, 12*(1), 8–18.

Alene, K. A., et al. (2019). Spatial patterns of tuberculosis and HIV co-infection in Ethiopia. *PLoS One, 14*(12), e0226127.

© The Author(s), under exclusive license to Springer Nature
Singapore Pte Ltd. 2023
M. Nishi, *Curing Lives*,
https://doi.org/10.1007/978-981-99-1831-7

Asfaw, D. (1958). Ekub. *Ethnological Society Bulletin* (University College of Addis Ababa) (8), 63–76.

Assefa, Y., et al. (2017). Performance of the antiretroviral treatment program in Ethiopia, 2005–2015: Strengths and weaknesses toward ending AIDS. *International Journal of Infectious Diseases, 60*, 70–76.

Bahru, Z. (2002a). *A history of modern Ethiopia, 1855–1991* (2nd ed.). Addis Ababa University Press.

Bahru, Z. (2002b). Systems of local governance among the Gurage: The Yejoka Qicha and the Gordana Sera. In Z. Bahru & S. Pausewang (Eds.), *Ethiopia· The challenge of democracy from below* (pp. 17–28). Nordiska Afrikainstitutet; Forum for Social Studies.

Balabanova, D., et al. (2013). Good health at low cost 25 years on: Lessons for the future of health systems strengthening. *The Lancet, 381*(9883), 2118–2133.

Bergström, S., & Goodburn, E. (2001). The role of traditional birth attendants in the reduction of maternal mortality. *Studies in Health Service Organization and Policy, 17*, 77–95.

Berhanu, Z. (2006). *Care and support and people living with HIV and AIDS at Holy water: An assessment at four selected sites in Addis Ababa* (Master thesis submitted to the Graduate School of Social work). Addis Ababa University.

Berhanu, Z. (2010). Holy Water as an Intervention for HIV/AIDS in Ethiopia. *Journal of HIV/AIDS and Social Services, 9*(3), 240–260.

Berhe, T., Gemechu, H., & Waal, A. D. E. (2005). War and HIV prevalence: Evidence from Tigray, Ethiopia. *African Security Review, 14*(3), 107–114.

Bhalotra, S. (2007). Spending to save? State health expenditure and infant mortality in India. *Health Economics, 16*(9), 911–928.

Biehl, J. (2004). The activist state: Global pharmaceuticals, AIDS, and citizenship in Brazil. *Social Text, 22*(3), 105–132.

Biehl, J. (2005). *Vita: Life in a zone of social abandonment.* University of California Press.

Biehl, J. (2007). Pharmaceuticalization: AIDS treatment and global health politics. *Anthropological Quarterly, 80*(4), 1083–1126.

Biehl, J., Good, B., & Kleinman, A. (2007). Introduction: Rethinking subjectivity. In J. Biehl, A. Kleinman, & B. Good (Eds.), *Subjectivity: Ethnographic investigations* (pp. 1–23). University of California Press.

Biehl, J., & Locke, P. (2017). Introduction: Ethnographic sensorium. In J. Biehl & P. Locke (Eds.), *Unfinished: The anthropology of becoming* (pp. 1–38). Duke University Press.

Bilal, N. (2012). Health extension program: An innovative solution to public health challenges of Ethiopia, a case study. USAID Health Finance and Governance Project. https://www.hfgproject.org/wp-content/uploads/2015/02/Health-Extension-Program-An-Innovative-Solution-to-Public-Health-Challenges-of-Ethiopia-A-Case-Study.pdf. Accessed April 20, 2020.

Birdthistle, I., et al. (2019). Recent levels and trends in HIV incidence rates among adolescent girls and young women in ten high-prevalence African countries: A systematic review and meta-analysis. *The Lancet Global Health, 7*(11), e1521–e1540.

Borrell, J. S., et al. (2020). Enset-based agricultural systems in Ethiopia: A systematic review of production trends, agronomy, processing and the wider food security applications of a neglected banana relative. *Plants, People, Planet, 2*(3), 212–228.

Breckenridge, A. (2003). Royal society of tropical medicine and hygiene meeting at the University of Liverpool, Liverpool, 16 March 2001 Debate that "This house believes the essential drug concept hinders the effective deployment of drugs in developing countries." *Transactions of the Royal Society of Tropical Medicine and Hygiene, 97*(1), 1.

Breckenridge, K. (2014). *Biometric state: The global politics of identification and surveillance in South Africa, 1850 to the present.* Cambridge University Press.

Brehony, E. (2010). Report on review of MMM Counselling and Social Services Centre, Addis Ababa, Ethiopia. *Medical Missionaries of Mary.* http://www.mmmworldwide.org/images/stories/pdf2010/mmm_cssc_review_2010_adis_ababa.pdf. Accessed June 10, 2019.

Brown, W. (2015). *Undoing the demos: Neoliberalism's stealth revolution.* Zone Books.

Central Statistical Agency. (2010). *Population and housing census of 2007: Report for southern nations, nationalities and peoples' region, part 1: Population size and characteristics.* Central Statistical Agency, Federal Democratic Republic of Ethiopia.

Chirac, P. (2003). Translating the essential drugs concept into the context of the year 2000. *Transactions of the Royal Society of Tropical Medicine and Hygiene, 97*(1), 10–12.

Cohen, M. S., & Gay, C. L. (2010). Treatment to prevent transmission of HIV-1. *Clinical Infectious Diseases, 50*(Suppl 3), S85–S95.

Cohen, M. S., et al. (2011). Prevention of HIV-1 infection with early antiretroviral therapy. *The New England Journal of Medicine, 365*(6), 493–505.

Development Assistance Group. (2013). *Intermediary INGO operations and the 70/30 guideline.* Development Assistance Group Ethiopia.

DiStefano, A. S., & Cayetano, R. T. (2011). Health care and social service providers' observations on the intersection of HIV/AIDS and violence among their clients and patients. *Qualitative Health Research, 21*(7), 884–899.

Duggal, R. (2007). Healthcare in India: Changing the financing strategy. *Social Policy & Administration, 41*(4), 386–394.

Eisinger, R. W., Dieffenbach, C. W., & Fauci, A. S. (2019). HIV viral load and transmissibility of HIV infection: Undetectable equals untransmittable. *JAMA, 321*(5), 451–452.

Ejigu, Y., & Tadesse, B. (2018). HIV testing during pregnancy for prevention of mother-to-child transmission of HIV in Ethiopia. *PLoS One, 13*(8), e0201886.

Epstein, H. (2007). *The invisible cure: Africa, the West, and the fight against AIDS*. Picador.

Ethiopian Public Health Institute. (2020). *Ethiopia population-based HIV impact assessment 2017–2018: Final report*. Ethiopian Public Health Institute.

Farmer, P. (1997). On suffering and structural violence: A view from below. In A. Kleinman, V. Das, & M. Lock (Eds.), *Social suffering* (pp. 261–283). University of California Press.

Fassin, D. (2007). *When bodies remember: Experiences and politics of AIDS in South Africa*. University of California Press.

Fassin, D. (2009). Another politics of life is possible. *Theory, Culture and Society, 26*(5), 44–60.

Federal Democratic Republic of Ethiopia. (2000). Revised Family Code Proclamation No. 213/2000, Federal Negarit Gazetta Extra Ordinary Issue 6/1, 1–92.

Federal Democratic Republic of Ethiopia. (2009). Charities and Societies Proclamation No. 621/2009, Federal Negarit Gazeta 15/25, 4521–4567.

Federal HIV/AIDS Prevention and Control Office. (2006). *Report on progress towards implementation of the declaration of commitment on HIV/AIDS*. Federal HIV/AIDS Prevention and Control Office, Federal Democratic Republic of Ethiopia.

Federal HIV/AIDS Prevention and Control Office. (2013). *Ethiopian national AIDS spending assessment report EFY 2004, 2011/12*. Federal HIV/AIDS Prevention and Control Office, Federal Democratic Republic of Ethiopia.

Federal Ministry of Health. (2005). *Health sector development plan III*. Federal Ministry of Health, Federal Democratic Republic of Ethiopia.

Federal Ministry of Health. (2015). *Health sector transformation plan.* Federal Ministry of Health, Federal Democratic Republic of Ethiopia.

Federal Ministry of Health. (2018). *National consolidated guidelines for comprehensive HIV prevention, care and treatment.* Federal Ministry of Health, Federal Democratic Republic of Ethiopia.

Federal Ministry of Women, Children and Youth Affairs. (2013). *National strategy and action plan on harmful traditional practices (HTPs) against women and children in Ethiopia.* Federal Ministry of Women, Children and Youth Affairs, Federal Democratic Republic of Ethiopia.

Ferguson, J. (1990). *The anti-politics machine: Development, depoliticization, and bureaucratic power in Lesotho.* Cambridge University Press.

Ferguson, J. (2015). *Give a man a fish: Reflections on the new politics of distribution.* Duke University Press.

Feyasa, M. B., Gebre, M. N., & Dadi, T. K. (2022). Levels of HIV/AIDS stigma and associated factors among sexually active Ethiopians: Analysis of 2016 Ethiopian demographic and health survey data. *BMC Public Health, 22*(1), 1080.

Friedland, G. H. (2000). Breaking the silence. *AIDS Clinical Care, 12*(8), 63–69.

Gannett, P., et al. (1989). Correspondence. *New England Journal of Medicine, 320*(17), 1150.

Garnett, G. P., et al. (2002). Antiretroviral therapy to treat and prevent HIV/AIDS in resource-poor settings. *Nature Medicine, 8*(7), 651–654.

Gaudilliere, J. -P., & Sunder Rajan, K. (2021). Making valuable health: Pharmaceuticals, global capital and alternative political economies. *BioSocieties, 16*(3), 313–322.

Geissler, P. W., & Prince, R. J. (2010). *The land is dying: Contingency, creativity and conflict in western Kenya.* Berghahn Books.

Getachew, T. (1990). Ethiopia: The course is charted. In E. Tarimo & A. L. Creese (Eds.), *Achieving health for all by the year 2000: Midway reports of country experiences* (pp. 80–97). World Health Organization.

Global Fund. (n.d.). *Country coordinating mechanism.* https://www.theglobal fund.org/en/country-coordinating-mechanism/. Accessed on November 19, 2022.

Goodburn, E. A., et al. (2000). Training traditional birth attendants in clean delivery does not prevent postpartum infection. *Health Policy and Planning, 15*(4), 394–399.

Graboyes, M. (2015). *The experiment must continue: Medical research and ethics in East Africa, 1940–2014.* Ohio University Press.

Granich, R. M., et al. (2009). Universal voluntary HIV testing with immediate antiretroviral therapy as a strategy for elimination of HIV transmission: A mathematical model. *The Lancet, 373*(9657), 48–57.

Greene, J. A. (2011). Making medicines essential: The emergent centrality of pharmaceuticals in global health. *BioSocieties, 6*(1), 10–33.

Guevara, M. W. (2006). Holy water cures even HIV. *The International Consortium of Investigative Journalists.* https://www.icij.org/investigations/divine-int ervention/holy-water-cures-even-hiv/. Accessed February 6, 2022.

Haraway, D. (1990). Reading Buchi Emecheta: Contests for women's experience in women's studies. *Women: A Cultural Review, 1*(3), 240–255.

Harries, A. D., et al. (2001). Preventing antiretroviral anarchy in sub-Saharan Africa. *The Lancet, 358*(9279), 410–414.

Heimer, C. A. (2007). Old inequalities, new disease: HIV/AIDS in sub-Saharan Africa. *Annual Review of Sociology, 33*(1), 551–577.

Herbst, J. (1996). Responding to state failure in Africa. *International Security, 21*(3), 120–144.

Hodes, R. M., & Kloos, H. (1988). Health and medical care in Ethiopia. *New England Journal of Medicine, 319*(14), 918–924.

Horton, R. (2000). African AIDS beyond Mbeki: Tripping into anarchy. *The Lancet, 356*(9241), 1541–1542.

Iliffe, J. (2006). *The African AIDS epidemic: A history.* James Currey.

International Crisis Group. (2003). *Ethiopia and Eritrea: War or peace? Nairobi.* International Crisis Group.

Izumi, K. (2007). Gender-based violence and property grabbing in Africa: A denial of women's liberty and security. *Gender and Development, 15*(1), 11–23.

Jiang, C., et al. (2020). Distinct viral reservoirs in individuals with spontaneous control of HIV-1. *Nature, 585*(7824), 261–267.

Jonsen, A. R. (1986). Bentham in a box: Technology assessment and health care allocation. *The Journal of Law, Medicine and Ethics, 14*(3–4), 172–174.

Kaleeba, N., & Ray, S. (2002). *We miss you all.* SAfAIDS.

Kalofonos, I. (2021). *All I eat is medicine: Going hungry in Mozambique's AIDS economy.* University of California Press.

Kenworthy, N., Thomann, M., & Parker, R. (2018). From a global crisis to the "end of AIDS": New epidemics of signification. *Global Public Health, 13*(8), 960–971.

Klausner, R. D., et al. (2003). Enhanced: The need for a global HIV vaccine enterprise. *Science, 300*(5628), 2036–2039.

Kleinman, A. (2006). *What really matters: Living a moral life amidst uncertainty and danger.* Oxford University Press.

Kloos, H. (1998). Primary health care in Ethiopia under three political systems: Community participation in a war-torn society. *Social Science & Medicine, 46*(4), 505–522.

Kloos, H., & Mariam, D. H. (2000). HIV/AIDS in Ethiopia: An overview. *Northeast African Studies, 7*(1), 13–40.

Kloos, H., et al. (2007). Utilization of antiretroviral treatment in Ethiopia between February and December 2006: Spatial, temporal, and demographic patterns. *International Journal of Health Geographics, 6*, 45.

Kloos, H., et al. (2013). Traditional medicine and HIV/AIDS in Ethiopia: Herbal medicine and faith healing, A review. *Ethiopian Journal of Health Development, 27*(2), 141–155.

Lakew, Y., Benedict, S., & Haile, D. (2015). Social determinants of HIV infection, hotspot areas and subpopulation groups in Ethiopia: Evidence from the National demographic and health survey in 2011. *British Medical Journal Open, 5*(11), e008669.

Leckie, J. R. et al. (2019). *Protecting land tenure security of women and girls in Ethiopia: Evidence from the land investment for transformation programme.* United Nations Economic Commission for Africa.

Lidman, M. (2016). In Ethiopia, a little round house is the strongest medicine against women's labor complications (Global Sisters Report). https://www.globalsistersreport.org/news/ministry/ethiopia-little-round-house-strongest-medicine-against-fistula-40896. Accessed on March 13, 2021.

Low-Beer, D., & Stoneburner, R. L. (2003). Behaviour and communication change in reducing HIV: Is Uganda unique? *African Journal of AIDS Research, 2*(1), 9–21.

Maes, K. (2016). *The lives of community health workers: Local labor and global health in urban Ethiopia.* Routledge.

Maes, K., & Kalofonos, I. (2013). Becoming and remaining community health workers: Perspectives from Ethiopia and Mozambique. *Social Science & Medicine, 87*, 52–59.

Mahajan, M. (2018). Rethinking prevention: Shifting conceptualizations of evidence and intervention in South Africa's AIDS epidemic. *BioSocieties, 13*(1), 148–169.

Marseille, E., Hofmann, P. B., & Kahn, J. G. (2002). HIV prevention before HAART in sub-Saharan Africa. *The Lancet, 359*(9320), 1851–1856.

Mazrui, A. A. (1995). The blood of experience: The failed state and political collapse in Africa. *World Policy Journal, 12*(1), 28–34.

McClellan, C. W. (1986). Coffee in centre–periphery relations: Gedeo in the early twentieth century. In D. L. Donham & W. James (Eds.), *The Southern marches of imperial Ethiopia: Essays in history and social anthropology* (pp. 175–195). Cambridge University Press.

McKay, R. (2018). *Medicine in the meantime: The work of care in Mozambique.* Duke University Press.

Mekuria, B. (1973). *Eder: Its roles in development and social change in Ethiopian urban centers* (4th-year senior essay, School of Social Work). Haile Sellassie I University.

Mills, E. J., et al. (2006). Adherence to antiretroviral therapy in sub-Saharan Africa and North America: A meta-analysis. *Journal of the American Medical Association, 296*(6), 679–690.

Mookherji, S., Ski, S., & Huntington, D. (2015). Tracking Global Fund HIV/AIDS resources used for sexual and reproductive health service integration: Case study from Ethiopia. *Globalization and Health, 11*(1), 21.

Moyer, E. (2015). The Anthropology of life after AIDS: Epistemological continuities in the age of antiretroviral treatment. *Annual Review of Anthropology, 44*(1), 259–275.

Murphy, M. (2017). *The economization of life.* Duke University Press.

Muthusamy, N., Levine, T. R., & Weber, R. (2009). Scaring the already scared: Some problems with HIV/AIDS fear appeals in Namibia. *Journal of Communication, 59*(2), 317–344.

Network of Networks of HIV Positives in Ethiopia. (2013). *Global Fund Round 7 Project implementation evaluation report.* Network of Networks of HIV Positives in Ethiopia.

Network of Networks of HIV Positives in Ethiopia. (n.d.). *Strategic plan 2008–2012.* Network of Networks of HIV Positives in Ethiopia.

Nguyen, V.-K. (2005). Antiretroviral globalism, biopolitics, and therapeutic citizenship. In A. Ong & S. J. Collier (Eds.), *Global assemblages: Technology, politic, and ethics as anthropological problems* (pp. 124–144). Blackwell.

Nguyen, V.-K. (2010). *The republic of therapy: Triage and sovereignty in West Africa's time of AIDS.* Duke University Press.

Nguyen, V.-K. (2015). Treating to prevent HIV: Population trials and experimental societies. In P. W. Geissler (Ed.), *Para-states and medical science: Making African global health* (pp. 47–77). Duke University Press.

Nishi, M. (2005). Making and unmaking of the nation-state and ethnicity in Modern Ethiopia: A study on the history of Silte people. *African Study Monographs Suppl., 29*, 157–168.

Nishi, M. (2008). Community-based rural development and the politics of redistribution: The experience of the Gurage Road Construction Organization in Ethiopia. *Nilo-Ethiopian Studies, 12*, 13–25.

Nishi, M. (2012). Morality, risk, and knowledge in the era of global health: HIV interventions and local responses among the Gurage. In *Paper presented at the 18th International Conference of Ethiopian Studies, Dire Dawa University, Dire Dawa, Ethiopia, October 28–November 3, 2012.*

Nyazema, N. Z., et al. (2000). Antiretrovial (ARV) drug utilisation in Harare. *The Central African Journal of Medicine, 46*(4), 89–93.

Nyblade, L., et al. (2003). *Disentangling HIV and AIDS stigma in Ethiopia, Tanzania and Zambia.* International Center for Research on Women.

Osseo-Asare, A. D. (2014). *Bitter roots: The search for healing plants in Africa.* University of Chicago Press.

Parkhurst, J. O. (2002). The Ugandan success story? Evidence and claims of HIV-1 prevention. *The Lancet, 360*(9326), 78–80.

Pfeiffer, J., & Chapman, R. (2015). An anthropology of aid in Africa. *The Lancet, 385*(9983), 2144–2145.

Pfeiffer, J., & Chapman, R. R. (2019). NGOs, austerity, and universal health coverage in Mozambique. *Globalization and Health, 15*(Suppl 1), 0.

Pfizenmaier, K. (2015). HIV Prevention and Counseling at Holy Water Sites in Ethiopia. *University of Washington Department of Global Health.* https://globalhealth.washington.edu/news/2015/09/28/hiv-prevention-and-counseling-holy-water-sites-ethiopia. Accessed April 2, 2022.

Piot, P. (2012). *No time to lose: A life in pursuit of deadly viruses.* W. W. Norton.

Piot, P., et al. (2008). Coming to terms with complexity: A call to action for HIV prevention. *The Lancet, 372*(9641), 845–859.

Piot, P., Zewdie, D., & Türmen, T. (2002). HIV/AIDS prevention and treatment. *The Lancet, 360*(9326), 86.

Popp, D., & Fisher, J. D. (2002). First, do no harm: A call for emphasizing adherence and HIV prevention interventions in active antiretroviral therapy programs in the developing world. *AIDS, 16*(4), 676–678.

President's Commission for the Study of Ethical Problems in Medicine and Biomedical and Behavioral Research. (1983). *Securing access to health care: A report on the ethical implications of differences in availability of health services, Volume one: Report.* President's Commission for the Study of Ethical Problems in Medicine and Biomedical and Behavioral Research.

Prince, R. (2012). HIV and the moral economy of survival in an East African City. *Medical Anthropology Quarterly, 26*(4), 534–556.

Provisional Military Administration Council. (1975). Government Ownership of Urban Lands and Extra Houses Proclamation No. 47/1975, Negarit Gazeta 34/41, 200–214.

Ramjee, G., & Daniels, B. (2013). Women and HIV in sub-Saharan Africa. *AIDS Research and Therapy, 10*(1), 30.

Rawls, J. (1999). *A theory of justice* (Revised). Belknap Press of Harvard University Press.

Rheinberger, H.-J. (1997). *Toward a history of epistemic things: Synthesizing proteins in the test tube.* Stanford University Press.

Rhodes, T., et al. (2019). The social life of HIV care: On the making of 'care beyond the virus.' *BioSocieties, 14*(3), 321–344.

Rose, N. (2006). *The politics of life itself.* Princeton University Press.

Rosen, L. (2003). *The culture of Islam: Changing aspects of contemporary Muslim life.* University of Chicago Press.

Sahle, M., Saito, O., & Demissew, S. (2021). Exploring the multiple contributions of enset (Ensete ventricosum) for sustainable management of home garden agroforestry system in Ethiopia. *Current Research in Environmental Sustainability, 3*, 100101.

Salim, S. A. K., & Karim, Q. A. (2001). Breaking the silence, one year later: Reflections on the Durban Conference. *AIDS Clinical Care, 13*(7), 63–65.

Serieux, J. E., et al. (2012). The Impact of the global economic crisis on HIV and AIDS Programs in a high prevalence country: The case of Malawi. *World Development, 40*(3), 501–515.

Shack, W. A. (1966). *The Gurage: A people of the ensete culture.* The International African Institute by Oxford University Press.

Shack, W. A., & Marcos, H.-M. (1974). *Gods and heroes: Oral traditions of the Gurage of Ethiopia.* Clarendon Press.

Simoni, J. M., et al. (2006). Self-report measures of antiretroviral therapy adherence: A review with recommendations for HIV research and clinical management. *AIDS and Behavior, 10*(3), 227–245.

Simpson, A. (2009). *Boys to men in the shadow of AIDS: Masculinities and HIV risk in Zambia.* Palgrave Macmillan.

Smith, K., et al. (2011). HIV-1 treatment as prevention: The good, the bad, and the challenges. *Current Opinion in HIV and AIDS, 6*(4), 315–325.

Sontag, S. (1989). *AIDS and its metaphors.* Farrar, Straus and Giroux.

Sori, A. T. (2012). Poverty, sexual experience and HIV vulnerability risks: Evidence from Addis Ababa, Ethiopia. *Journal of Biosocial Science, 44*(6), 677–701.

Stevens, W., Kaye, S., & Corrah, T. (2004). Antiretroviral therapy in Africa. *British Medical Journal, 328*(7434), 280–282.

Stommes, E., & Sisaye, S. (1980). The development and distribution of health care services in Ethiopia: A preliminary review. *Canadian Journal of African Studies, 13*(3), 487–495.

Swaminathan, H., et al. (2008). *Women's property rights, HIV and AIDS, and domestic violence: Research findings from two districts in South Africa and Uganda.* HSRC Press.

Teshome, T. (2002). Reflections on the Revised Family Code of 2000 Ethiopia. In: A. Bainham (Ed.), *International survey of family law 2002 edition* (pp. 153–170). Family Law.

Tilley, H. (2011). *Africa as a living laboratory: Empire, development, and the problem of scientific knowledge, 1870–1950.* University of Chicago Press.

Tura, H. (2014). *Women's right to and control over rural land in Ethiopia: The law and the practice.* Social Science Research Network.

UK Government. (2014). Guidance on the US Board on Geographic Names (BGN)/Permanent Committee on Geographical Names (PCGN) romanization systems. https://www.gov.uk/government/publications/romanization-systems. Accessed January 11, 2023.

Vergne, L., et al. (2002). Resistance to antiretroviral treatment in Gabon: Need for implementation of guidelines on antiretroviral therapy use and HIV-1 drug resistance monitoring in developing countries. *Journal of Acquired Immune Deficiency Syndromes, 29*(2), 165–168.

Wang, H., et al. (2016). *Ethiopia Health Extension Program: An institutionalized community approach for universal health coverage.* The World Bank.

WHO Global Programme on AIDS. (1995). *Taso Uganda: The inside story, participatory evaluation of HIV/AIDS counselling, medical and social services, 1993–1994.* World Health Organization.

Whyte, S. R. (2014a). Introduction: The first generation. In S. R. Whyte (Ed.), *Second chances: Surviving AIDS in Uganda* (pp. 1–24). Duke University Press.

Whyte, S. R. (Ed.). (2014b). *Second chances: Surviving AIDS in Uganda.* Duke University Press.

Whyte, S., van der Geest, S., & Hardon, A. (Eds.). (2009). *Social lives of medicines.* Cambridge University Press.

Woldemeskel, G. (1989). The consequences of resettlement in Ethiopia. *African Affairs, 88*(352), 359–374.

World Bank. (2000). *Can Africa claim the 21st century?* The World Bank.

World Health Organization. (1977). *The selection of essential drugs* (WHO Technical Report Series no. 614). World Health Organization.

World Health Organization. (1992). *Traditional birth attendants: A joint WHO/UNFPA/UNICEF statement*. World Health Organization.

World Health Organization. (2002a). *Global action for skilled attendants for pregnant women*. World Health Organization.

World Health Organization. (2002b). *Scaling up antiretroviral therapy in resource-limited settings*. World Health Organization.

World Health Organization. (2017). Reaching key populations to prevent the spread of disease in Ethiopia. *WHO Regional Office for Africa*. https://www.afro.who.int/news/reaching-key-populations-prevent-spread-disease-ethiopia. Accessed April 2, 2022.

Woubshet, D. (2015). *The calendar of loss: Race, sexuality, and mourning in the early era of AIDS*. Johns Hopkins University Press.

Wubshet, M., et al. (2013). Death and seeking alternative therapy largely accounted for lost to follow-up of patients on ART in northwest Ethiopia: A community tracking survey. *PLoS One, 8*(3), e59197.

Yamey, G., & Rankin, W. W. (2002). AIDS and global justice. *British Medical Journal, 324*(7331), 181–182.

YeGurage Hizib Ras Ages Limat Dirigit. (2007). *Qicha: YeGurage bahilawi hig, teshashilo yewetta (Revised edition of the Gurage customary law) [2000EC]*. Finfine Matemiya na Publishing.

Yeltyopya Kwankwawoch Tinat na Mirmir Maikel. (2001). *Amerigna mezgebe kalat* (Amharic dictionary) [1993EC]. Artistik Matemiya Dirijit.

Zuger, A. (2012). A skeptic looks at 'test and treat.' *Journal Watch HIV/AIDS Clinical Care*. https://search.proquest.com/scholarly-journals/skeptic-looks-at-test-treat/docview/1319247515/se-2?accountid=11929. Accessed on March 13, 2021.

Index

© The Author(s), under exclusive license to Springer Nature Singapore Pte Ltd. 2023
M. Nishi, *Curing Lives*,
https://doi.org/10.1007/978-981-99-1831-7